ETHNOMATHEMATICS
A Multicultural View of Mathematical Ideas

ETHNOMATHEMATICS
A Multicultural View of Mathematical Ideas

Marcia Ascher

CHAPMAN & HALL/CRC

Boca Raton   London   New York   Washington, D.C.

Published in 1991 by
Chapman & Hall/CRC
Taylor & Francis Group
6000 Broken Sound Parkway NW, Suite 300
Boca Raton, FL 33487-2742

© 1991 by Taylor & Francis Group, LLC
Chapman & Hall/CRC is an imprint of Taylor & Francis Group

No claim to original U.S. Government works
Printed in the United States of America on acid-free paper
15 14 13 12 11 10 9 8 7

International Standard Book Number-10: 0-412-98941-7 (Softcover)
International Standard Book Number-13: 978-0-412-98941-4 (Softcover)
Library of Congress catalog number: 90-48677

Library of Congress Cataloging-in-Publication Data

Ascher, Marcia, [date]
Ethnomathematics : a multicultural view of mathematical ideas / Marcia Ascher.
   p. cm.
Includes bibliographical references and indexes.
ISBN 0-412-98941-7
1. Ethnomathematics. I. Title.
GN476.15.A83 1991
510—dc20   90-48677

Visit the Taylor & Francis Web site at
http://www.taylorandfrancis.com

and the CRC Press Web site at
http://www.crcpress.com

To Bob

a c k n o w l e d g m e n t s

My interest in the mathematical ideas of traditional peoples began when I collaborated with an anthropologist in a study of an Inca artifact that was said to be somehow numerical. That study became more extensive than I had anticipated, and, as part of it, I had to rethink many of my ideas about mathematics and about the relationship of mathematics and culture. I also became increasingly conscious of the omission or misrepresentation of traditional peoples in the mathematics literature. My additional investigations led to the creation of a college-level course on the subject. Teaching this course over the years required that I refine my thoughts and develop means of conveying them to others. I am indebted to Claudia Zaslavsky for making me aware of other scholars who were concerned with somewhat similar issues. The Wenner-Gren Foundation provided partial support for the study of the Inca artifact, discussed in Chapter 1, and for my study of graphs in cultures, discussed in Chapter 2.

The specific impetus to write this book was a conference at the Mathematical Research Institute at Oberwolfach; there I found historians of mathematics who welcomed new ideas. An invitation to spend the academic year 1987–1988 as a Getty Scholar at the Getty Center for the History of Art and the Humanities in Santa Monica, California, provided the opportunity to begin writing. The administration and staff of the Center made every effort to make the year valuable and special. My particular thanks to Carole Frick and Lori Repetti (then graduate students) for their assistance and sincere interest in my work. Ithaca College provided the reduction of one course so that I could complete the book.

Although they are also acknowledged in the text, I thank those who permitted me to reproduce figures or photographs. My appreciation also goes to all of those who helped by typing, reading parts of the manuscript, suggesting further sources, and questioning and commenting on what I wrote or what I said. I am especially grateful to the following reviewers of the manuscript for their helpful comments: H.J.M. Bos of the University of Utrecht, The Netherlands; Paul Campbell of Beloit College; Donald W. Crowe of the University of Wisconsin; Ubiratan D'Ambrosio of UNICAMP, Brazil; James Rauff of Millikin University; Melanie Schneider of the University of Wisconsin–Madison; Alvin White of Harvey Mudd College; and Claudia Zaslavsky.

Above all, I thank my husband, Robert Ascher, for his encouragement, critical reading, and constructive suggestions.

Marcia Ascher

c o n t e n t s

i n t r o d u c t i o n

Let us take a step toward a global, multicultural view of mathematics. To do this, we will introduce the mathematical ideas of people who have generally been excluded from discussions of mathematics. The people are those who live in traditional or small-scale cultures; that is, they are, by and large, the indigenous people of the places that were "discovered" and colonized by Europeans.

The study of the mathematical ideas of traditional peoples is part of a new endeavor called *ethnomathematics*. Mathematicians and others are usually skeptical of newly coined fields, wondering if they have any substance. To answer this justifiable concern, we begin with quite specific mathematical ideas as they are expressed and embedded in some traditional cultures. Some of the peoples whose ideas are included are the Inuit, Navajo, and Iroquois of North America; the Incas of South America; the Malekula, Warlpiri, Maori, and Caroline Islanders of Oceania; and the Tshokwe, Bushoong, and Kpelle of Africa. Only afterward will you find a discussion of the scope and implications of ethnomathematics and how it relates to other areas of inquiry.

The concept of *culture* is subtle and multifaceted; thus, to capture its essence, the word has numerous definitions and elaborations. What most have in common, and what is significant for us, is that in any culture the people share a language; a place; traditions; and ways of organizing, interpreting, conceptualizing, and giving meaning to their physical and social worlds. Because of the spread of a few dominant cultures, there is no culture that is completely self-contained or unmodified. The effects that were wrought or things that were adopted or adapted concern us only peripherally. We focus, rather, on what is called *the ethnographic present*, that is, the period during which the traditional culture holds full sway.

As is the case for well over 95 percent of all cultures, until very recently each of the cultures we draw upon had no system of writing. As a result, there are no early records *by them* in their *own* words. Most of our information is based on the writings of others who translated what they heard and what they observed into their own terms. For Europeans, the fifteenth and sixteenth centuries were a time of exploration. They discovered then, and continued to discover, that there were many people in the world living in ways quite different from their own. Explorers, traders, and missionaries wrote about the people they encountered but, because these people were so different from themselves, the descriptions vary considerably in reliability. During the late nineteenth century, but particularly during this century, more meaningful descriptions and greater understanding became available with the growth of the fields of ethnology, culture history, and linguistics. From twentieth-century insights, theories, and knowledge, we have come to understand that there is no single linear path along which cultures progress, with some ahead and others behind. Cultures share some ideas but not others. Even where an idea is the same or similar, it will be differently expressed and have different contexts in different cultures. This is as true for mathematical ideas as it is for other ideas; the Western expression is but one of many.

Among mathematical ideas, we include those involving number, logic, spatial configuration, and, even more significant, the combination or organization of these into systems or structures. Our interest is this broad realm of mathematical ideas. *Mathematics* has no generally agreed-upon definition; it means to some whatever was included in their school or college courses and to others whatever is done by the Western professional class called mathematicians. Some attempts to define mathematics emphasize its objects of study and others its methods; some definitions are extremely narrow and others exceptionally vague and broad. In general, however, concerns about what mathematics *is* are generally in the domain of philosophers and the historians who write its history. Their opinions have changed through time as new lines of inquiry developed, as earlier assumptions were reexamined, and, most particularly, as the ensemble of beliefs in the world around them changed. In any case, their definitions of mathematics are based solely on the Western experience even though they are often phrased universally. As a result, the category "mathematics" is Western and so is not to be found in traditional cultures.

That is not to say that the ideas or concepts we deem mathematical do not exist in other cultures; it is rather that others do not distinguish them and class them together as we do. The same is true for many of our categories; art, work, and entertainment are but a few other examples that have no direct analogue in other cultures. Indeed, how people categorize things is one of the major differences between one culture and another. So, to avoid being constrained by the Western connotations of the word mathematics, we speak, instead, about mathematical ideas. The particular ideas, the way they are expressed, the context and ideational complex of which they are a part, vary depending on the culture. The contexts could be, for example, what Westerners designate as art, navigation, religion, record keeping, games, or kin relations. These are, in fact, some of the contexts for the ideas we discuss. Further, our discussion places these ideas within their proper cultural and ideational contexts so that they remain the ideas and expressions of others and do not become just pale reflections of our own.

Nonetheless, despite viewing the mathematical ideas of others in their contexts, we must keep in mind that we are limited by our own mathematical and cultural frameworks. It is more likely that we can see or understand those ideas that are in some way similar to our own, while ideas that that we do not in some way share may escape us. Moreover, as we try to discuss the ideas of others, we will, of necessity, recast them into our Western mode. And, at times, in trying to convey the significance of ideas, we will do so by elaborating on our Western expressions of them. Throughout, we differentiate—and trust that the reader will do so as well—between mathematical ideas that are implicit and those that are explicit, and between Western concepts that we use to describe or explain and those concepts we attribute to people in other cultures.

c h a p t e r

Numbers:
Words and
Symbols

o   n   e

Counting numbers are often the first association we make with things mathematical, and so that is where we shall begin. At the start, a strong distinction must be made between spoken words and symbols. Our written words, *one, two, three,* and so on, are symbols that represent the sounds of the spoken words; we also represent numbers by the written symbols 1, 2, 3. . . . We say the word *two* when we encounter the letter combination *t-w-o* or the symbol 2, but we do not need these symbols to know or use the spoken word. For now we are discussing only spoken words; later we will discuss symbols as well.

Basically the number concept is the recognition of a single entity combined with the understanding that another can be added to it, another then added to that aggregate, another to that, and so on. The *and so on* is crucial—the process can be extended indefinitely. Number words are simply the names given to the series that is so formed. The capacity to count is a human universal related to human language. It reflects an extremely important property that distinguishes human language from other animal communication systems. The property is *discrete infinity*; that is, humans create sentences that can contain an unlimited number of discrete words. Different cultures make different use of the counting facility; some generate lots of number words and a very few generate almost none. How many words are generated does not reflect differences in capability or understanding; it reflects the degree of number concern in the culture.

In Western culture there is belief that numbers carry a great deal of information. The belief has become particularly pervasive over the last hundred years. We are asked almost daily to answer some question of "How many?" But we are often uneasy that the numerical answer will not really convey the essence of what should be conveyed. "How many bedrooms in your house?" "Two, but one is so small that..." or "How many hours a day do you work?" "Well, it depends on..." An increase in the uniformity of material goods has been concomitant with the growth in viewing many things as standardized units rather than as intrinsically unique. Thus, the number of TV sets in the United States, the number of hamburgers sold in a fast-food chain, the number of highway deaths are the kinds of facts that bombard us daily. Our belief in the objectivity of numerical statements is so strong that we associate numbers with human intelligence via IQ's, college readiness via SAT scores, and even happiness or satisfaction via other scores. Most other cultures have less belief in the value of the information conveyed by numbers. This difference in concern may be associated with technology or population size or the domination of other interests such as spirituality, aesthetics, or human relations. A few native Australian groups, for example, are noted for their lack of interest in numbers. They are also noted for the richness of their spiritual life and overriding focus on human relations. Perhaps most important, however, is that many other cultures value contextual understanding rather than decontextualization and objectivity.

The conclusion that the counting facility is intrinsic to being human is based on both contemporary linguistic theory and accumulated evidence. It resolves a question that had caused much speculation among mathematicians. The question was how numbers originated, whether they were invented or discovered, and, in either case, by whom. Philosophers of Western mathematics, concerned with clarifying the foundations of mathematics, debated whether the "natural" numbers (positive integers) could be assumed or whether they needed definition. Those who defined them did so in ways that directly imply that (a) "one" is a natural number and (b) if some "x" is a natural number, so is its successor. This formal version reiterates our earlier statement: among the natural numbers there is some entity (we call it one); since one is a natural number so is one plus one (we call it two); then since two is a natural number, so is two plus one (we call it three), and so is its successor (we call it four), and so on. Another philosophical school, the Intuitionists, found it inappropriate to give a definition of the natural numbers because their construction is indeed natural to all human beings.

3 When creating number words, a culture could, and some few do, give each number an independent name. However, the set of number words of a culture is generally patterned and has implicit within it arithmetic relationships.

Consider our set of numerals. It basically consists of root words for one, two, three through nine, ten, hundred, thousand, and million. These are then used and reused in a cyclic pattern based on a cycle length of ten: one through ten, then eleven through twenty, twenty-one through thirty, and so on until ten such cycles reach one hundred. From there to two hundred, the ten cycles of ten are repeated. Ten repetitions of this larger cycle reach to one thousand, and then one thousand of these still larger cycles reach to one million. Thus there are cycles of ten within cycles of hundred, within cycles of thousand, within cycles of million. Further, the number words within the cycles reflect implicit multiplication and addition. For example, implied in seventy-three is (seven times ten) plus three and in four hundred and forty seven is (four times one hundred) plus (four times ten) plus seven.

In the diverse sets of numeral words from other cultures there are, of course, different sets of root words. Although many numeral systems have some cyclic pattern, the basic cycle lengths vary. And there are even differences in the implied arithmetic operations. The cyclic constructions underscore the open-ended, extendable nature of numbers, and the implicit operations reinforce that they are from an interrelated set. As an example, consider Nahuatl numerals. Nahuatl is a language of Central Mexico. There are root words for what we would symbolize as 1 through

5, 10, 15, 20, 400, and 8000. The numerals have a cyclic pattern based on cycles of length twenty. Just as our cycle based on ten has a root word for ten times ten (hundred), so this pattern has a root word for twenty times twenty (four hundred) and for twenty of those (eight thousand). Within the cycle, there are reference points at five, ten, and fifteen. The implied arithmetic operations are addition and multiplication. Combining these, and using our symbols only as shorthand for their words, the Nahuatl number words can be described as:

1	11 implies 10 + 1
2	12 implies 10 + 2
3	13 implies 10 + 3
4	14 implies 10 + 4
5	15
6 implies 5 + 1	16 implies 15 + 1
7 implies 5 + 2	17 implies 15 + 2
8 implies 5 + 3	18 implies 15 + 3
9 implies 5 + 4	19 implies 15 + 4
10	20

21 implies 20 + 1

36 implies 20 + (15 + 1)

41 implies (2 × 20) + 1

56 implies (2 × 20) + (15 + 1)

72 implies (3 × 20) + (10 + 2)

104 implies (5 × 20) + 4

221 implies [(10 + 1) × 20] + 1

400

463 implies 400 + [(3 × 20) + 3]

7463 implies [(15 + 3) × 400] + [(13 × 20) + 3]

8000

8861 implies 8000 + (2 × 400) + (3 × 20) + 1.

Another language with a cyclic pattern based on twenty is Chol, a Mayan language spoken in northern Chiapas, Mexico. This, too, has a reference point at ten, but none at five or fifteen. The implied operations are also addition and multiplication. However, above twenty, in one of its numeral forms, the combination of these elements is distinctly different and is sometimes referred to as overcounting: a numeral is related to the next higher cycle rather than the preceding one. The root words in the set are one through ten, twenty, four hundred, and eight thousand. Using ten as a reference point, the words for eleven through nineteen imply (ten plus one), (ten plus two), (ten plus three), ... (ten plus nine). Then, for example:

21 implies 1 toward (2×20)

36 implies $(10 + 6)$ toward (2×20)

500 implies (5×20) toward (2×400)

564 implies $400 + [4$ toward $(9 \times 20)]$

1055 implies $(2 \times 400) + [(10 + 5)$ toward $(10 + 3) \times 20]$

8861 implies $8000 + (2 \times 400) + [1$ toward $(4 \times 20)]$.

Nahuatl and Chol numerals both have cycles based on twenty, but languages with number word cycles based on four, five, eight, and ten are also reportedly common, some with and some without additional reference points within the cycles. Yet other languages have other types of number word patterns. One such example is Toba (western South America) in which the word with value five implies (two plus three), six implies (two times three), and seven implies (two times three) plus one. Then eight implies (two times four), nine implies (two times four) plus one, and ten is (two times four) plus two. There is also implied subtraction in, for example, some Athapaskan languages on the Pacific Coast. In them, eight implies (ten minus two) and nine implies (ten minus one).

The foregoing illustrations have been drawn from Native American languages. They do reflect, however, the diversity found throughout the world. For all of them, it is important to keep in mind that number words are historically derived; they are not formulas and none are any more sensible or necessary than others. There is an often-repeated idea that numerals involving cycles based on ten are somehow more logical because of human fingers. The Yuki of California are said to believe that their cycles based on eight are most appropriate for *exactly* the same reason. The Yuki, however, are referring to the interfinger spaces. For each people, their own number words are simple, obvious, and spoken without recourse to arithmetic or calculation. And, to emphasize that number words constructed differently from our own are not confined to non-literate peoples, we conclude with two European examples. In French number words, there are cycles based on ten, but, within them, starting at sixty, a cycle based on twenty intrudes. Again, using symbols as shorthand for their words:

60 implies (6 × 10), ..., 69 implies (6 × 10) + 9, but then

70 implies (6 × 10) + 10

71 implies (6 × 10) + 11

72 implies (6 × 10) + 12

$\vdots$

80 implies (4 × 20)

81 implies (4 × 20) + 1

82 implies (4 × 20) + 2

$\vdots$

99 implies (4 × 20) + (10 + 9).

And, in Danish,

20 implies (2 × 10)

30 implies (3 × 10)

40 implies (4 × 10)

50 implies $(3 - \frac{1}{2}) \times 20$

60 implies (3 × 20)

70 implies $(4 - \frac{1}{2}) \times 20$

80 implies (4 × 20)

90 implies $(5 - \frac{1}{2}) \times 20$.

 Another aspect of language intimately connected with number words is numeral classifiers. These occur in a considerable number of the world's languages including Chinese, Japanese, almost all of the languages in Southeast Asia, numerous languages in Oceania, some African languages, and a few European languages such as Gaelic and Hungarian. Numeral classifiers are terms that are included when number words are spoken with nouns. They convey information about the nouns that is qualitative rather than quantitative but, nevertheless, a *necessary* part of making quantitative statements. The number of different classifiers within a language varies considerably. Some have as few as two classifications; others have as many as two hundred.

A simple illustration is the language of the Maori, the indigenous people of New Zealand. When a statement is made about a number of human beings, it must contain the classifier for humans. It is as if we were to say "five humans women" or "five humans Californians." No other numerical statement has a classifier. The distinction between humans and everything else or between the animate and inanimate is common to almost all languages using classifiers. Earlier we noted that many other peoples have more concern for context. In this use of classifiers, there is an insistence that the significant characteristic of life be carried along with any statement of number.

As yet there is no unanimity on classifiers. Some linguists believe they are counterparts of measures used with continuals, such as water. We cannot *count* water without introducing something discrete with which to measure it, such as "five cups of water" or "five gallons of water." Or it may be that the classifiers are sets to which the particular objects belong or that they prepare the hearer for qualities of the objects being counted. But, in any case, the classifiers are a link between quantity and quality. In some languages, classifiers are used only with small numbers; in some languages, they are not used with ten, hundred, thousand, or whatever is comparable depending on the cycle size; and in Thai, for example, the presence of a classifier with large numbers implies an actual count was made, whereas its absence implies estimation or approximation.

In the example of the Maori, a clear distinction between classes was possible: human versus everything else. For most languages such statements are not possible. Lists can be made of nouns that need a particular classifier and sometimes it can be stated, more or less, what characteristics emerge from these nouns as a group. These descriptions give insight into how others categorize the world, but they do not lead to generalizations that match our conceptual categories. Beyond the association of nouns and their characteristics with the classifiers in a particular language, there are commonalities across languages. The most basic distinction, as we noted previously, is animate versus inanimate and, within that, human versus nonhuman. Sometimes humans are further classified by either social rank or kinship. For inanimate things and sometimes even nonhuman animate things, shape is often involved. Distinctions are based on dimensionality (long versus flat versus spherical) and size (small versus large) combined with rigidity versus flexibility. These distinctions, of course, introduce another idea of mathematical importance, namely that *geometric* properties are particularly significant in these classificatory schemes.

For some specific examples, we introduce two languages from Micronesia with numeral classifiers that are few or moderate in number. Gilbertese is spoken on the Gilbert Islands, now part of the Republic of

Kiribati, where people live primarily by fishing and by eating coconuts, pandanus, breadfruit, and taro. While their classifiers reflect some of the general ideas mentioned previously (animacy, human groups, flat things, round things, elongated things), they also reflect the specifics of their environment. The classifiers, denoted here by (a)–(r), are:

a. animates (except for fish longer or larger than people), spirits, and ghosts
b. groups of humans, especially family groups
c. days
d. years
e. generations
f. coconut thatch
g. bundles of thatch
h. rows of thatch
i. rows of things (except of thatch)
j. layers
k. pandanus fruit
l. elongated things (posts, bones, dried pandanus fruit stored in long tubular containers, fish larger or longer than the speaker, fingers, sharks)
m. leaves, pages of books, playing cards
n. baskets, dances
o. customs
p. all modes of transportation
q. things from which edibles are detached (trees, plants, shrubs, fish-hooks, land sections)
r. general (often replacing other classifiers of inanimates and used for serial counting).

Since the Gilbertese classifiers become affixed to the number words, it would be as if we were to say "sixp ships," "sixl fingers," or "sixr houses." Kusaiean, another Micronesian language, has just two classifications, which also combine into the number words. The two are (a) four-legged animals, insects, fish, forms of transportation, and elongated or pointed objects—including rivers or roads; and (b) all other nouns, including humans, fruits, nets, atolls, trays, and all ordinals.

A more elaborate classification scheme is outlined in Chart 1.1. The fifty-five numeral classifiers of the Dioi language (spoken in Kweichow Province, S. China) are denoted in the chart as:

(a), ..., (z), (aa), ..., (zz), (aaa), (bbb), (ccc).

As nonspeakers of the language, we may not see the classes as mutually exclusive or exhaustive, and we cannot easily understand or summarize them. But, most important, as you read through the chart, keep in mind that these are distinctions for numerical statements. Five (k) roses conveys flowers as well as the number five: in Dioi the five alone is insufficient and improper.

Chart 1.1. Dioi numeral classifiers

a. Debts, credits, accounts
b. Mountains, walls, territory
c. Legal processes, legal affairs
d. Forensic affairs, tribunal affairs
e. Opium pipes, whistles, etc.
f. Sores, wounds, blows
g. Rice fields
h. Showers, storms
i. Nets, mesh
j. Letters, packages
k. Flowers
l. Debts
m. Clothing, covers, bedclothes
n. Pieces of cloth
o. Potions, medicines
p. Children, sons, etc.
q. Plants on the stalk, trees
r. Shavings, etc.
s. Pairs of things
t. Feasts, rounds of drinks
u. Skins, mats, carpets, peels/coatings, covers
v. Books as volumes, books of one volume
w. Books composed of multiple volumes, account books, registers
x. Books in one volume, single volumes of a book, books, registers
y. Old people, government men, persons whom one respects
z. Persons, spirits, men, angels, God, workers, thieves
aa. Girls, young women
bb. Sheets of paper and of other materials, napkins, planks; titles to property; accusations and other writings
cc. Flat stones, planks, fields, large rocks, cakes of glutinous rice
dd. Routes, rivers, cords, carrying poles, chains
ee. Aspects, features of things, characteristics, traits
ff. Things strung on threads or rods as meat, beads
gg. Pieces of meat, bones, keys, locks, combs, brushes, feather dusters
hh. Certain flat things such as planks, beams, wood shavings, tiles, rice cakes
ii. Children, small pieces of money, small stones and pebbles
jj. Pieces of things and things generally spherical, such as heads, stones, fists, money, mountains
kk. Common objects such as houses, boats, furniture, plates and dishes, firearms; abstracts
ll. Living beings including birds, fishes, animals, friends, persons, thunder, rural spirits
mm. Things in clusters such as grapes, strings of mushrooms, collars of bells for horses

Chart 1.1 (*cont'd*)

nn. Objects that come in pairs and are rarely separated, such as shoes, arms, chopsticks

oo. Things that have handles or arms, such as knives, hoes, many tools

pp. Large beams, boards, hewn stones

qq. Certain things that come in pairs or tens, such as services of bowls, parallel inscriptions

rr. Speech phrases, sentences, events, affairs

ss. Long, rigid things such as poles, sticks, pipes, columns

tt. Covers and bedclothes, drapes, linens

uu. Pieces of money

vv. Some objects such as swing plows, harrows, looms

ww. Needles, pins, and other sharply pointed objects

xx. Certain flat, very thin things such as sheets of paper, large leaves, coattails, flaps of clothing

yy. Grains, pills, things in granular form, shot for a gun, buttons, drops of rain

zz. Skeins or hanks, bobbins of thread, tied handfuls of vermicelli, small sticks of incense

aaa. Things strung on lines or rods; fish on strings, meat on skewers

bbb. Objects that present a flat surface, slices of meat, sheets of paper or cloth, fields, flat cakes of glutinous rice, dresses/clothing

ccc. Hands and feet and objects of similar form, such as ginger root, columns in rows in front of a house, honeycombs, etc.

An early misunderstanding of numeral classifiers became translated into a significant mathematical misunderstanding of nonliterate peoples. The theory of classical evolution, a late-nineteenth-century/early-twentieth-century paradigm, assumed that there was a single linear evolutionary path leading from savagery to civilization in a series of predestined stages. What appeared to be different number words for different things were taken to be from earlier stages before the general concept of number was understood. To this day, many histories of mathematics associate numeral classifiers only with nonliterate people and repeat that peoples using them do not understand that *two* human fishermen and *two* long things bananas both contain the same concept of *two*. From this, these historians concluded that such people are incapable of abstraction. But it is clear that numeral classifiers combined into or coupled with number words do not interfere with the concept of number. It would be no less abstract and no less understanding of seven hundred if one had to say seven hundred *human* traffic fatalities rather than seven hundred traffic fatalities. We would not belabor the point if the implications of this misunderstanding were not so mathematically crucial and pervasive as well as persistent. There is ample evidence that it is time to move ahead.

6 At the start of this chapter, we made a strong distinction between spoken number words and number symbols. Now, as we discuss number symbols, we must make a further distinction between symbols that are marks conveying speech sounds and those that are not. When we see, for example, our number symbol *8*, we associate with it our spoken word *eight*. If, instead, we were French, we would associate the same symbol with the spoken word *huit*. Different words can be used because the number symbol does not represent the sound of the number word; it represents the concept behind the word. Furthermore, the shape and form of the symbol could be different and yet still represent the same concept and, hence, be associated with the same word. *VIII*, for example, is also a symbol for which we say *eight*. The written letter combination *e-i-g-h-t* conveys to us the same meaning. That letter combination, however, *does* represent the sound of our number word as contrasted to *8* and *VIII*, which do not.

The distinction that we are drawing is between recorded number symbols and written words. Not all peoples had recorded number symbols just as not all peoples had writing. The two, however, are often so closely linked that their differences are overlooked. We will look at the recorded number symbols used by the Incas. It is a particularly interesting example because the Incas had no writing; all of the records they did have were encoded on spatial arrays of colored knotted cords called *quipus*. Our earlier examples were brief, and this will be our first extended one with more about the mathematical idea involved and more about the cultural complex in which it is embedded.

Numbers played a significant and extensive role in the Inca quipus. They were used not only for quantities but also as labels. Consider the use in our culture of telephone numbers, for example, the telephone number 207-276-5531. The three digits 207 are an area code; they identify that the telephone is in the state of Maine. Then 276 stands for a region within that area, namely Mt. Desert Island, and, finally, 5531 specifies a particular phone on that island. A quantity answers a question such as "How much?" or "How many?" or "How long?" A label is simply an identifier that could just as well be letters or any familiar shapes. It makes sense to do arithmetic with numbers that are quantities. But, on the other hand, to double a phone number and add three would have no meaning. There are increasingly many number labels in the world around us—your social security number, the ISBN number of this book, the product codes on most everything that is purchased, and so on. The more information we process by computers, the more such number labels are used. Part of the power of computers is that they can electronically represent, store, and process both quantities and labels. The quipus are records, not calculating or processing devices. The combined use of numbers as quantities and numbers as labels made the quipus sufficiently flexible to serve as the only record-keeping system of a large, complex, bureaucratic state.

How these quantities and labels were organized and what more is known about the logical-numerical system of the quipus is another longer story. Here we confine ourselves to how numbers, whether they be quantities or labels, were symbolized. Because the medium is so different from our own, some rudiments of the quipu's construction and logic are needed. But first we shall introduce the Incas and the important role that quipus played in their communication network.

The Incas, as we use the term, were a complex culture of three to five million people that existed from approximately 1400 to 1560 A.D. The region they inhabited is described today as all of the country of Peru and portions of Ecuador, Bolivia, Chile, and Argentina. The terrain varies from coastal deserts to tropical forests to the mountainous highlands of the Andes. Many different groups were in the region but, starting about 1400, one of the groups, the Incas, moved slowly and steadily upon the others. The groups retained many of their individual traditions but were forcibly consolidated into a single bureaucratic entity. Using but extending various aspects of the cultures of the different groups, the Incas achieved consolidation by the overlay of a common state religion and one common language, Quechua; the broad extension of a road system and irrigation systems; the imposition of a system of taxation involving agricultural products, labor, finished products such as cloth, and even lice if that was all one had; the relocation of groups of people; and the building and use of storehouses to hold and redistribute agricultural products as well as to feed the army as it moved. Within about thirty years after the Europeans reached the Andes and "discovered" the Incas, the Inca culture, in essence, was destroyed.

Because the Incas had no writing, we have no information about them in their own words. What information we do have is fragmentary: some comes from the chronicles of Europeans who were a part of the very group that justified its destructive acts on the basis of cultural superiority. Some is obtained from the study of Inca artifacts that survived, and among these artifacts are the quipus. In Inca practice, people were buried with objects they had made or used while alive. All of about 500 quipus, that today are housed in museums through the world, were recovered from gravesites in relatively dry coastal areas.

The Incas can be characterized as methodical, highly organized, concerned with detail, and intensive data users. The Inca bureaucracy continuously monitored the areas under its control. They received many messages and sent many instructions daily. The messages included details of resources such as items that were needed or available in storehouses, taxes owed or collected, census information, the output of mines, or the composition of work forces. The messages were transmitted rapidly using the extensive road system via a simple, but effective, system of runners. A runner carried a message from the posthouse where he was stationed to the next posthouse on the road, where a new runner stood ready to take the message the next few miles to another posthouse, and so on. The messages had to be clear, compact, and portable. Quipumakers were responsible for encoding and decoding the information. In each of the occupied regions, the Incas selected a few people with some local standing and sent them to Cuzco, the capital, to learn Inca ways in general and quipumaking in particular. The people then returned home to serve as bureaucrats in the Inca administration. In addition to dealing with resources, the messages on quipus, according to the Spanish chroniclers, were as varied as ballads, peace negotiations, laws, and state history.

A quipu is an assemblage of colored knotted cotton cords. Among the Inca, cotton cloth and cordage were of great importance. Used to construct bridges, in ceremonies, for tribute, and in every phase of the life cycle from birth to death, cotton cordage and cloth were of unparalleled importance in Inca culture and, hence, not a surprising choice for its principal medium. The colors of the cords, the way the cords are connected together, the relative placement of the cords, the spaces between the cords, the types of knots on the individual cords, and the relative placement of the knots are all part of the logical-numerical recording. Figure 1.1 shows a quipu in the rolled form in which it was probably transported. Figure 1.2 shows the same quipu unrolled. Just as anything from simple scratch marks to complex mathematical notation can be contained on a piece of paper or a blackboard, the logical-numerical system embedded in the cord arrays sets them quite apart from any other knotted cord usage by individuals or groups in other cultures. The quipus are distinctively Inca and unique to them.

Figure 1.1. A quipu that is completed and rolled (courtesy of the Smithsonian National Museum, Washington, D.C.)

Figure 1.2. The quipu of Figure 1.1 unrolled

In general, a quipu has one cord, called the *main cord*, that is thicker than the rest and from which other cords are suspended. When the main cord is laid horizontally on a flat table, most of the cords fall in one direction; these are called *pendant cords*. Those few that fall in the opposite direction are called *top cords*. Suspended from some or all of the pendant or top cords are other cords called *subsidiary cords*, and there can be subsidiaries of subsidiaries, and so on. Sometimes a single cord is attached to the end of the main cord. Because it is attached in a different way from the pendant or top cords, it is referred to as a *dangle end cord*. All the attachments are tight so that once the quipu is constructed, the cord positions are fixed. A schematic of a quipu is shown in Figure 1.3. In that illustration, by spacing along the main cord, the pendants are formed into two different groups. And the first cord has two subsidiaries on the same level, while the fourth cord has subsidiaries on two different levels. Pendant cords, top cords, and subsidiary cords range from 20 to 50 centimeters in length. A quipu can be made up of as few as three cords or as many as 2000 cords and can have some or all of the cord types described.

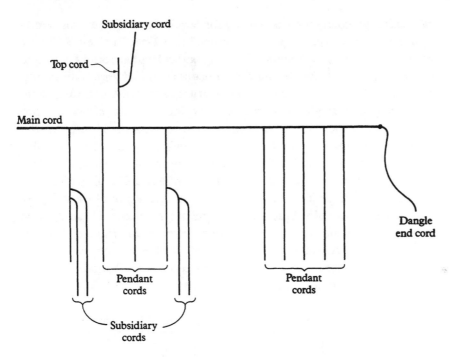

Figure 1.3. A schematic of a quipu

The colors of the cords are important. A cord can be made up of several strands of a single color or a combination of differently colored strands. An entire quipu can be made up of cords of a single color or can have as many as 50 or 60 different colors or color combinations. Just as spacing along the main cord associates some cords and distinguishes them from others, color also associates and distinguishes cords. A color—say, red—did not have a specific universal meaning; it marked, within the context of a particular quipu or set of quipus, a relationship to other red cords and differences from, say, blue cords.

The type and level of the cords, the relative spacing of the cords, and the colors of the cords were used to create the logic of a quipu array. The knots on the cords then represent the numbers appropriately placed into this specific arrangement. Consider, for example, a record that we might keep of the number of potatoes consumed by four different families, broken down within each family by men, women, and children. We might arrange these, as in Figure 1.4, into four columns of three rows each. If we were using a quipu rather than a sheet of paper, we might use spacing to form four groups of three cords each where each group contains the data for one family, and, within each group, the numbers for the men, women, and children are on the first, second, and third pendants, respectively. Or, instead, the families might be distinguished from each other by color, such as red (*R*) for family 1, yellow (*Y*) for family 2, and blue (*B*) and green (*G*) for families 3 and 4 (see Figure 1.5). There are several other assortments of spacing and color that could be used to organize these data. But one particular use of color had important ramifications for the Inca mode of number representation. Continuing with our potato consumption example, let the colors red (*R*), yellow (*Y*), and blue (*B*) be associated with men, women, and children, respectively. Then each consecutive repetition of the color pattern red-yellow-blue would be another family. The significance of this technique is that pendants can be omitted from a group without ambiguity when the category with which it is associated is inapplicable. As you look at Figure 1.6, note that the color blue has been omitted for the third family; that is, there will be no potato consumption recorded for children because family 3 has no children. This elimination of "blanks" makes for compactness but, more crucial, it clarifies that a cord that is present with no knots has some other meaning.

	Family 1	Family 2	Family 3	Family 4
Men				
Women				
Children				

Figure 1.4. Layout of potato consumption record

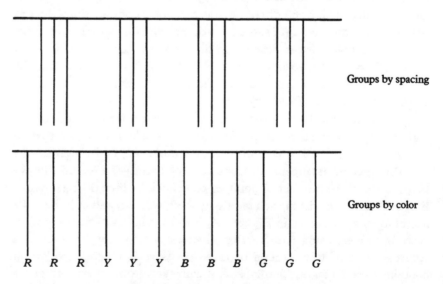

Groups by spacing

Groups by color

R *R* *R* *Y* *Y* *Y* *B* *B* *B* *G* *G* *G*

Figure 1.5. Quipu layouts of potato consumption record

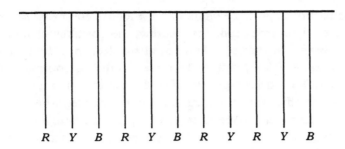

Figure 1.6. Another quipu layout of the potato consumption record

Number representation on quipus is of an exceptionally sophisticated type. In fact, although the symbols are made up of knots rather than marks on paper, it is basically the same as ours. It is a base 10 positional system. Although our familiarity with our own system makes it seem most obvious, it is but one of many possible systems and its development took about 4000 years. To clarify what is meant by our base 10 positional system, consider the meaning of 2308. In our system, there are just 10 individual symbols: 0, 1, 2, 3, 4, 5, 6, 7, 8, 9. To express any value, we select from these and place them in some order. The value of the collection depends on where the individual symbols are placed. The collection has the value $(2 \times 1000) + (3 \times 100) + (0 \times 10) + (8 \times 1)$. Each consecutive position, moving to the left, is multiplied by 10 another time. The number of times 10 is multiplied is called the *power* of ten, and so each consecutive position is one higher power of 10. The concept of a base positional system is not tied to the base 10; any positive integer above 1 could be used just as well. For example, if the base is 7, only 7 individual symbols are needed. The value of any collection of these depends on where each symbol is placed but, in base 7, each consecutive position, moving to the left, is one higher power of 7. With this positional rule and using our individual symbols 0, 1, 2, 3, 4, 5, 6, any value can be represented. The collection 6505 interpreted in the base 7 positional system means $(6 \times 343) + (5 \times 49) + (0 \times 7) + (5 \times 1)$. It has the same value as what was symbolized in the base 10 as 2308.

To fully appreciate the significance of a base positional system, consider the system of Roman numerals, which is not of this type. It was one of the systems in use in Europe through the seventeenth century and is still in use today in some contexts in Western culture. The individual symbols are I, V, X, L, C, D, and M. In our base 10 notation, the values of these are 1, 5, 10, 50, 100, 500, and 1000, respectively. Although the value of a symbol collection depends on where the symbols are placed, it is the relationship of the value of a symbol to the value of nearby symbols that is of importance rather than the positions of individual symbols per se. For example, the value of a collection is the sum of the individual values *if* the symbols decrease in value as you move to the right. Thus, XVI translates into our 16 because it is interpreted as $10 + 5 + 1$. On the other hand, subtraction is indicated when a symbol of smaller value precedes one of larger value; IX translates into our 9. In the collection MCMXLIV, both the additive and subtractive principles are involved. A direct translation of each symbol in this collection is 1000, 100, 1000, 10, 50, 1, 5. Since the 100 precedes the 1000, it is subtracted from it as is 10 from 50 and 1 from 5, but these differences decrease in value and so their results are added. In all, the collection is translated as:

$$1000 + (1000 - 100) + (50 - 10) + (5 - 1) = 1944$$

The formation of a symbol collection is less clear as there are no complete explicit rules. For example, the value written in our system as 80 could be written in Roman numerals as XXC (interpreted as $100 - 10 - 10$) or as LXXX (interpreted as $50 + 10 + 10 + 10$). LXXX is used because, in general, only a single symbol of lesser value immediately precedes a symbol of greater value.

An important contrast between a base positional system and the Roman numerals is that, in a base positional system, no new symbols are needed as numbers increase in magnitude. However, with the Roman numeral symbols introduced so far, large numbers would be difficult to represent. A number with value as large as our 120,000 would have required a collection of 120 consecutive M's. Although they are rarely seen now, additional symbols were used to indicate thousandfold. One such form was a horizontal stroke above a number so that $\overline{\text{CXX}}$, for example, would be translated as: $1000(100 + 10 + 10) = 120,000$. Another contrast is that the size of Roman numeral representation is not related to the magnitude of the number. The single symbol M, for example, is greater in value than the collection XXXVIII, whereas in a base positional system the size of the representation and the value are always related.

Above all, however, the most important feature of any base positional system is that it simplifies and enables arithmetic. *Because there are a limited and specific set of symbols and explicit rules for forming them into other symbols, general arithmetic principles can be developed and explicitly stated.* Some historians of Western mathematics have even attributed Western progress in arithmetic and calculation to the advent of our base 10 positional system.

There are some similarities between the concepts underlying a base positional system and the cycles and arithmetic relationships that are seen in the number words of many cultures. For our own words, the cycle is of length 10, but, as we have noted, other cycle lengths are also used. Despite the similarity in our case, where number symbols do exist, the number words and number symbols need not correspond. The French, for example, imply (4 × 20 + 3) in *quatre-vingt-trois*, their word for their base 10 symbol 83. And, we still say one thousand nine hundred and forty-four, although we write MCMXLIV.

It should be clear now that, although the base 10 positional system is most familiar to us, the Inca use of it was neither necessary nor to have been expected. They might not have had a positional system at all, and even with a positional system, they need not have chosen the base 10.

On individual quipu cords, whether they are pendants, top cords, subsidiaries, or dangle end cords, there are only three types of knots: single knots (simple overhand knots), long knots (made up of two or more turns), and figure-eight knots (see Figure 1.7). Knots are clustered together and separated by space from adjacent clusters; each cluster contains from no knots to nine knots. Basically, the knot clusters represent the digits. The data can be one number or multiple numbers. Where the knots on a cord form one number, it is an integer in the base 10 positional system. From the free end of a cord to where it is attached to another cord, each consecutive cluster position is valued at one higher power of 10. The units position is further distinguished from the other positions by the type of knot used; it has long knots, while the other positions have clusters of single knots. Because of the way long knots are constructed, there cannot be a long knot of one turn, and so a figure-eight knot is used for a one in the units position. Where the knots on a cord represent multiple numbers, reading from the free end of the cord, each long knot (or figure-eight knot) is the units position of a new number.

The concept of zero is important in itself, but it is crucial to any base positional system. Our number 302, for example, has a value that is quite different from the values of 3002 or 32. Without a zero—that is, something standing for nothing—we could not know which was intended. In its entirety, the concept of zero has three aspects:

a. that "nothing" is identified in some way;
b. that a position can contain "nothing" yet contribute to the overall value of a number; and

Figure 1.7. The three types of knots

c. that "nothing" can stand alone and be treated as a number in and of itself.

On the quipus, zero is indicated by the absence of knots in a cluster position. Lack of a special symbol causes little ambiguity because other devices are used instead. The units position is clearly identified by knot type, and knot cluster positions are aligned from cord to cord so that a position with no knots is apparent when related to other cords. And the use of color patterning deals with the last aspect of the concept. The value zero is a number in and of itself; its representation is a cord with no knots. Were it not for the fact that color patterning enables the omission of blanks, the zero value would not be distinguishable from a cord left blank.

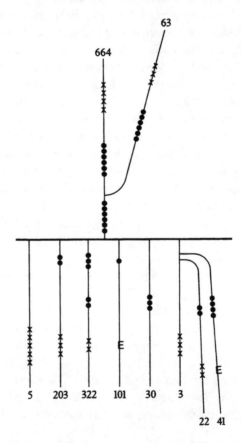

Figure 1.8. Numbers represented by knots (● = single knot, x = one turn of a long knot, E = figure-eight knot)

The difference is that in the example of the family potato consumption (Figure 1.6), after the numbers are encoded on the quipu, a blue cord with no knots would indicate that zero potatoes were consumed by the children, whereas the absence of a blue cord indicates that the family had no children.

Figure 1.8 is a schematic with examples of numbers represented by knots. Multiple numbers are shown in Figure 1.9. Since the highest position on each is closest to the point of attachment, the digits on top cords are read upward, while those on the pendants are read downward. For the subsidiaries, the reading depends on the direction of the cord to which they are attached. The main cord, pendant cords, spaces between cords, color differences, and knots and knot clusters are all visible in the photograph in Figure 1.10. The alignment of knot clusters from cord to

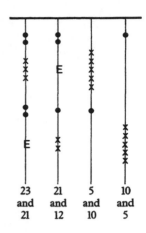

Figure 1.9. Multiple numbers represented by knots (● = single knot, x = one turn of a long knot, E = figure-eight knot)

Figure 1.10. A quipu in the collection of the Museo Nacional de Antropología y Arqueología, Lima, Peru

cord, which required careful planning and careful execution, can also be seen.

The Incas were what is generally termed a civilization except for one attribute: namely, they had no writing system. They had, nonetheless, this symbolic representation of numbers, which served as the cornerstone for recording information. Utilizing numbers, the logical-numerical system embedded in the quipus was sufficient to serve the needs of the complex, highly organized Inca state.

 Number words, whether a culture has few or many, have a variety of formations, and many cultures have numeral classifiers, while we do not. The Incas shared the belief in the importance of numerical information and shared the concept of a base 10 positional system. Their expression and elaboration of these, however, were quite different. We have also seen that some mathematical ideas, when they are present, pervade a culture, while others are limited to a specially trained or selected few. The same, of course, is also true for mathematical ideas in our culture.

Even things as familiar to us as number words and number symbols must be considered afresh when we look at other cultures. We cannot assume that they share all of our Western concepts, and, even when the concepts are shared, we cannot expect that they will be expressed in the same way.

Notes

1. This description of the counting capacity as a human universal follows Noam Chomsky, *Language and Problems of Knowledge: The Managua Lectures*, MIT Press, Cambridge, Mass., 1988, pp. 167–169, 183. Also see Burt W. Aginsky and Ethel G. Aginsky, "The importance of language universals," *Word*, 4 (1948), 168–172 and Kenneth Hale, "Gaps in grammar and culture" in *Linguistics and Anthropology in Honor of C. F. Voegelin*, M. D. Kinkade, K. L. Hale, and O. Werners, eds., The Peter de Ridder Press, Lisse, 1975, 195–316. James R. Hurford, *Language and Number*, Basil Blackwell, London, 1987 is a comprehensive discussion of how people use and acquire numerals. It relates linguistic analysis to psychological and social considerations, as well as addressing the philosophy of number. Hurford sees the human number faculty as innate but attributes it to different capacities than does Chomsky. The book is recommended for all the issues it confronts and its extensive bibliography. My usage of the word *numerals* follows Hurford: "Whatever numbers actually are, I will assume that numerals are used by people to name them (using 'name' in a non-technical sense, that is not implying that numerals are logically names, as opposed to, say, predicates or quantifiers)" (p. 7). Thus, a numeral system can be a system of words as well as a system of, say, graphic forms.

For a fascinating discussion of the growth of number usage in the United States see Patricia Cline Cohen, *A Calculating People: The Spread of Numeracy in Early America*, University of Chicago Press, Chicago, 1982. Stephen Jay Gould in *The Mismeasure of Man*, Norton, N.Y., 1981 discusses the attempts to assign numerical measures to human intelligence.

2. The different philosophical views are widely discussed in the mathematical literature. For a straightforward discussion, see Philip J. Davis and Reuben Hersh, *The Mathematical Experience*, Houghton Mifflin Co., Boston, 1981, pp. 393–411; and for a specific description of the Intuitionist view of natural numbers, see A. Heyting, *Intuitionism: An Introduction*, North-Holland Publishing Co., Amsterdam, 1956, pp. 13–15.

3. The Nahuatl words are from Stanley E. Payne and Michael P. Closs, "Aztec numbers and their uses" in *Native American Mathematics*, Michael P. Closs, ed., University of Texas Press, 1986, pp. 213–235; the Chol words from Wilbur Aulie, "High-layered numerals in Chol (Mayan)" *International Journal of American Linguistics*, 23 (1957) 281–283; the Toba words from W. J. McGee, "Primitive Numbers," *Smithsonian Institution, Bureau of American Ethnology, 19th Annual Report*, Washington, D.C., 1900, pp. 821–851; the Athapaskan words from Virginia D. Hymes, "Athapaskan numeral systems," *International Journal of American Linguistics*, 21 (1955), 26–45; and the Danish words from James R. Hurford, *The Linguistic Theory of Numerals*, Cambridge University Press, Cambridge, 1975. Collections of number words span about one hundred years and reflect large changes in understanding. Some earlier works such as W. C. Eells, "Number systems of the North American Indians," *American Mathematical Monthly*, 20 (1913) 263–299, are good reference sources, but theoretically outdated.

Although about a different part of the world, we particularly recommend John Harris, "Facts and fallacies of aboriginal number systems," *Work Papers of Summer Institute of Linguistics—Australian Aborigines Branch*, S. Hargrave, ed., series B, 8 (1982) 153–181, and his 1987 article "Australian Aboriginal and Islander mathematics," pp. 29–37 in *Australian Aboriginal Studies*, number 2. They contain significant observations and are important as a refutation of the frequently heard generalization that all Native Australians count only to three or four. Also, Jadran Mimica's book *Intimations of Infinity*, Berg Publishers Ltd., Oxford, 1988, is an important contribution to viewing number in cultural context. In the words of its author, "The book is an interpretation of the counting system and the meanings of the category of number among the Iqwaye people of Papua New Guinea." It explores "their intrinsic relations with the Iqwaye view of the cosmos" (p. 5).

Although it is not as generally applicable as the author suggests, the descriptive model proposed by Zdenek Salzmann, "A method for analyzing numerical systems," *Word* 6 (1950), 78–83 is recommended.

4. The discussion of numeral classifiers is based on the overviews found in "Universals, relativity, and language processing," Eve V. Clark & Herbert H. Clark, in *Universals of Human Language* vol. 1, Joseph H. Greenberg, ed. Stanford University Press, Stanford, Calif., 1978, pp. 225–277; Joseph H. Greenberg, "Numeral classifiers and substantival number: problems in the genesis of a linguistic type," *Proceedings of the Eleventh International Congress of Linguists in Bologna (1972)*, Luigi Heilmann, ed., Mulino, Bologna, 1974, pp. 17–38; Karen L. Adams and Nancy Faires Conklin, "Toward a theory of natural classification," *Papers of the ninth regional meeting of the Chicago Linguistic Society*, 9 (1973) 1–10; Karen Lee Adams, *Systems of Numeral Classification in the Mon-Khmer, Nicobarese, and Aslian Subfamilies of Austroasiatic*, Ph.D. thesis, University of Michigan 1982; and Nancy Faires Conklin, *The Semantics and Syntax of Numeral Classification in Thai and Austronesian*, Ph.D. thesis, University of Michigan, 1981; as well as the more specific articles, Robbins Burling, "How to choose a Burmese numerical classifier," *Context and Meaning in Cultural Anthropology*, Melford E. Spiro, ed., The Free Press, N.Y., 1965, 243–264; Mary R. Haas, "The use of numerical classifiers in Thai," *Language*, 18 (1942) 201–205; Kathryn C. Keller, "The Chontal (Maya) numeral system," *International Journal of American Linguistics*, 21 (1955) 258–275; and Martin G. Silverman, "Numeral-classifiers in the Gilbertese language," *Anthropology Tomorrow*, 7 (1962) 41–56.

We note in particular that the appendices of the thesis by Conklin contain the classifiers and associated nouns for 36 languages. Chart 1.1 is adapted from her Appendix A.

5. The discussion of Inca quipus is based on portions of *Code of the Quipu: A Study in Media, Mathematics, and Culture*, M. Ascher and R. Ascher, University of Michigan Press, Ann Arbor, 1981. Here the emphasis is on the representation of numbers; there the context of the quipus in Inca culture, the role of the quipumakers, and the overall logical-numerical system of the quipus are elaborated. The book contains numerous examples and extensive references. It is written for the general reader. For those who wish to study quipus in greater depth, there are detailed descriptions of 215 quipus in the Aschers' *Code of the Quipu: Databook* (1978) and *Code of the Quipu: Databook II* (1988) available on microfiche from Cornell University Archivist, Ithaca, NY. These descriptions

are based on firsthand study of specimens in museums and private collections in
North America, South America, and Europe. The *Databooks* also contain the
locations of all known quipus and a complete bibliography of previously published descriptions.

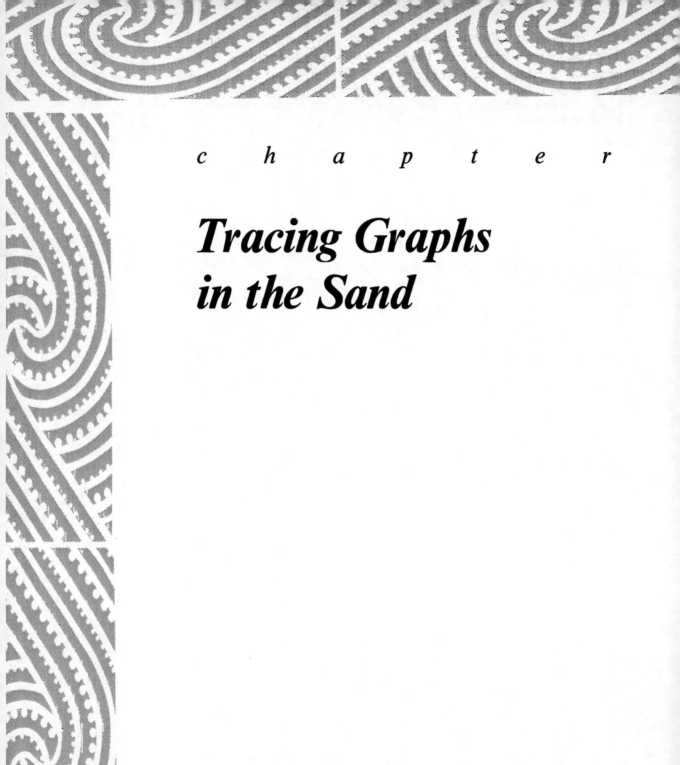

c   h   a   p   t   e   r

*Tracing Graphs
in the Sand*

t w o

In about 1905 a European ethnologist studying among the Bushoong in Africa was challenged to trace some figures in the sand, with the specification that each line be traced once and only once without lifting his finger from the ground. The ethnologist was unaware that a group of people in his own culture were keenly interested in such figure tracing. And, being unfamiliar with what Western mathematicians call graph theory, he was also unaware of how to meet the challenge.

The Bushoong who challenged the Western ethnologist were children concerned with three such figures. Among the Tshokwe, who live in the same region in Africa, sand tracing is not a children's game and hundreds of figures are involved. We will discuss both of these cases as well as a third from another part of the world, which is quite separate and more elaborate. Not only is each case from a different culture but their contexts within the cultures differ markedly. The mathematical idea of figures traced continuously is the central thread of this chapter. But for each culture there are, as well, other geometric and topological ideas involved in the creation, regularities, or relationships of these spatial forms. These too will be discussed, as our primary interest is mathematical ideas and, as with any ideas, they occur in complexes that do not necessarily conform to particular Western categories. To begin, however, a few ideas from graph theory need to be introduced.

Graph theory, described geometrically, is concerned with arrays of points (called *vertices*) interconnected by lines (called *edges*). This field has been growing in importance in our culture because it provides new approaches, stimulates new concepts, and has many applications. Graphs are particularly useful in studying flows through networks. For example, traffic flow involves intersections (considered to be the vertices) interconnected by roads (considered to be the edges).

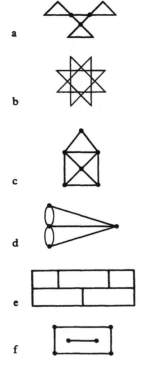

    a

    b

    c

    d

    e

    f

A few graphs are shown in Figure 2.1. In graph theoretic terminology, Figures 2.1a–2.1e are said to be *connected planar* graphs. A *connected* graph is one in which each vertex is joined to every other one via some set of edges. (In contrast to the others, Figure 2.1f is not connected.) A *planar* graph is one that lies entirely in the plane; that is, it need not be depicted as rising out of this flat paper. A freeway overpass and the road beneath it, for example, would not be represented by a planar figure.

A classical question in graph theory is: for a connected planar graph, can a continuous path be found that covers each edge once and only once? And, if there is such a path, can it end at the same vertex as it started? This is the question that is said to have inspired the beginnings of graph theory by the mathematician Leonhard Euler. According to the story, there were seven bridges in Königsberg (then in East Prussia), where Euler lived. The bridges spanned a forked river that separated the town into four land masses. The townspeople were interested in knowing if, on their Sunday walks, they could start from home, cross each bridge once and only once and end at home. Euler showed that for the particular situation such a route was impossible and also started considering the more general question. Between Euler in 1736 and Hierholzer some 130 years later, a complete answer was found. To state the result another term

Figure 2.1. Graphs

is needed, namely the *degree* of a vertex. The degree of a vertex is the number of edges emanating from it; a vertex is odd if its degree is odd and even if its degree is even. The answer to the question, first of all, is that not all connected planar graphs can be traced continuously covering each edge once and only once. If such a path can be found, it is called, in honor of Euler, an *Eulerian path*. Such a path exists if the graph has only one pair of odd vertices, provided that the path begins at one odd vertex and ends at the other. Also, such a path can be found if all of the vertices are even and, in this circumstance, the path can start from any vertex and end where it began. The graphs for which there cannot be Eulerian paths are those that have more than one pair of odd vertices.

With these results in mind, look again at the graphs in Figure 2.1. Graph a and graph b have all vertices of degree 4. Each can, therefore, be traced continuously covering every edge once and only once, beginning at any vertex and ending at the same place. Graph c has six vertices, two of degree 3 and four of degrees 2 or 4. It, therefore, has an Eulerian path that begins at one odd vertex and ends at the other. For graph d and graph e, no Eulerian paths exist as the former has four odd vertices and the latter eight.

Attempts to trace these five graphs are also of interest from a historical perspective. Graph d is a representation of the Königsberg bridge problem, with the vertices standing for the land masses and the edges standing for the bridges. Graphs a, b, c, and e link the domain of professional Western mathematicians to Western folk culture. Graph c may well be familiar to you, as tracing it is a children's street puzzle in many cities, from New York to London to Berlin. Graphs a, b, and e are from a collection of nineteenth-century Danish party puzzles. What is more, graph b, entitled "the nightmare cross," was said to have magical significance. Of these graphs, graph e is perhaps the most ubiquitous; not only was it a folk puzzle in nineteenth-century Denmark, but it has also appeared and reappeared in mathematical treatises since at least 1844. And the eminent philosopher Ludwig Wittgenstein, in discussing the very foundations of mathematics, used the problem of tracing a quite similar figure as one that captures the essence of the subject.

With this brief background, we return to the figure tracing challenge posed by the Bushoong children.

3   The Bushoong are one subgroup of the Kuba chiefdom. The chiefdom consists of at least four different ethnic groups separated into some fourteen subgroups. The hereditary Bushoong chief is the *nyimi*, the sovereign chief among all the chiefs, and so the Bushoong live in and around the Kuba capital (see Map 1). In the late nineteenth century, there were about 100,000 Kuba with about 4000 people living in the capital. Those living in the capital were the *nyimi*, hundreds of his wives, nobility, and specialized craftspeople. The Belgian government became the colonial authority just after 1910, ruling the Kuba territory indirectly until the establishment of Zaire in 1960.

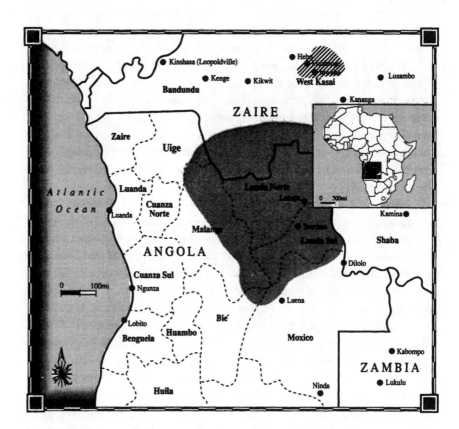

Map 1. The Angola/Zaire/Zambia region. The crosshatched region is inhabited by the Bushoong, and the Tshokwe are in the majority in the shaded area.

In the Kuba system of exchange the Bushoong have the role of decorators; in particular, they are sculptors of wood and embroiderers of cloth. Every third day a market is held in the capital and the Bushoong obtain pottery, salt, meat, fish, ivory, brass, wood, and plain cloth from the other subgroups while supplying embroidered cloth, sculpted wooden objects, masks, woven belts, hats, and raffia velour. Decoration of daily utilitarian objects as well as ceremonial objects is done for more than its intrinsic aesthetic value. Self-decoration and the possession of decorated objects bring prestige, and with prestige come political appointments and authority. Political position focuses around the *nyimi* and his appointed councils and titleholders; the former are men and the latter, both men and women. The relationship between decoration and political power is underscored by one element of the enthronement ceremony of the *nyimi*. At this ceremony he proclaims a special design that will be his official sign.

It is within this context, in a Bushoong village within the Kuba capital, that children play games drawing figures in the sand (see Figure 2.2). As with all Bushoong geometric decorations, each figure has a particular name. Naming, however, raises the very important point of cultural differences in perception. The name of a figure depends on how it is categorized. The Bushoong view a design as composed of different elementary designs, and the name given to a figure is the name associated with its most significant constituent. Thus, designs that appear the same to a Westerner may have different names, and those that appear different may have the same name. Moreover, significance is also related to process, and so sculptors may differ from embroiderers in the names they assign. Figure 2.2a is named *imbola*, which relates it to figures of the same name found carved on wooden objects, embroidered on cloth, and used as female scarification. In both name and appearance it is the same as a design formed of cowrie shells on a royal sash still in use in the 1970s. Figures 2.2b and 2.2c, named *ikuri banga* and *jaki na nyimi*, have no immediate counterparts in other figures but do look somewhat like some embroidered designs.

Figure 2.2a has seven vertices, each with four emanating edges. It can, therefore, be traced continuously starting anywhere and ending where one started. The starting and ending point used by the Bushoong children in their continuous tracing is marked by our label *S* on the figure. There is no record, however, of the specific tracing path that they used. When a continuous path is possible, there are many different ways that it can be accomplished. To emphasize the diversity, four possible paths are shown in Figure 2.3. Follow each of them with a pointer or your finger. You will see that Figure 2.3a first completes one of the triangular components and then the other, while Figure 2.3b goes around the exterior edges and then around the interior edges. Figures 2.3c and 2.3d, on the other hand, have no readily apparent plan. Other paths are also possible. How one "sees" the figure affects the path taken or, conversely, the path taken affects how one "sees" the figure.

Figures 2.2b and 2.2c each contain two vertices from which emanate an odd number of edges. From graph theory we know that tracing the figure continuously requires starting at one of the odd vertices and

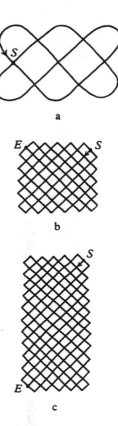

Figure 2.2. Bushoong sand figures. *S* and *E* denote the starting and ending points of the Bushoong tracings.

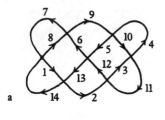

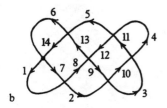

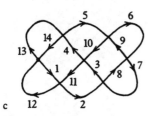

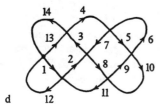

Figure 2.3. Some possible tracing paths

ending at the other. The Bushoong children, somehow, were aware of this as can be seen from their beginning and ending points (marked by our labels *S* and *E* on the figures). What is more, the tracing procedures they used for these two figures are known to us and we can see that they used a system and what that system was. For Figure 2.2b, starting at the uppermost right edge, proceed diagonally down to the left as far as possible, then up left as far as possible, up right as far as possible, then down right as far as possible, and so on going down left (dl), up left (ul), up right (ur), down right (dr). Because of the systematic nature of their drawing procedure, a concise set of instructions in terms of direction and number of units in that direction can be stated:

$$\text{dl } 10$$
$$\text{ul } 1, \text{ur } 10, \text{dr } 2, \text{dl } 9$$
$$\text{ul } 3, \text{ur } 8, \text{dr } 4, \text{dl } 7$$
$$\text{ul } 5, \text{ur } 6, \text{dr } 6, \text{dl } 5$$
$$\text{ul } 7, \text{ur } 4, \text{dr } 8, \text{dl } 3$$
$$\text{ul } 9, \text{ur } 2, \text{dr } 10, \text{dl } 1$$
$$\text{ul } 10.$$

By observation of the patterned number of units in each direction on successive sweeps, their procedure can be summarized as:

$$\text{dl } 10$$
$$\begin{cases} \text{for } i = 1, \text{ then } 3, \text{ then } 5, \text{ then } 7, \text{ then } 9, \\ \text{ul } i, \text{ur } 11 - i, \text{dr } i + 1, \text{dl } 10 - i \end{cases}$$
$$\text{ul } 10.$$

If we were to trace a similar figure where there are, in general, *N* units rather than 10 units along the first diagonal, *N* would have to be an even number. Then the instructions, stated in terms of this general *N*, would be:

$$\text{dl } N$$
$$\begin{cases} \text{for } i = 1, \text{ then } 3, \text{ then } 5, \text{ then } 2 \text{ more each time until } N - 1, \\ \text{ul } i, \text{ur } N + 1 - i, \text{dr } i + 1, \text{dl } N - i \end{cases}$$
$$\text{ul } N.$$

The tracing procedure used by the Bushoong for Figure 2.2c is a bit more complex. Again the tracing starts at the uppermost right edge and proceeds diagonally down to the left. As before, directions are changed after going as far up or down as possible. The procedure, in terms of number of units in each direction, is:

dl 10, dr 11

dl 1, ul 10, ur 11, ul 2, dl 9, dr 11

dl 3, ul 8, ur 11, ul 4, dl 7, dr 11

dl 5, ul 6, ur 11, ul 6, dl 5, dr 11

dl 7, ul 4, ur 11, ul 8, dl 3, dr 11

dl 9, ul 2, ur 11, ul 10, dl 1, dr 11

dl 10.

It is particularly noteworthy that both tracing procedures used by the Bushoong are systematic and the systems are variations on a like theme.

For the Tshokwe, the drawing of continuous figures in the sand is part of a widespread storytelling tradition. The figures, called *sona* (sing. *lusona*), are drawn exclusively by men. Because of the disintegration of native cultures brought about by colonialism, it is primarily older men who are knowledgeable and proficient in the drawing skill. The skill combines the memory of the drawings, the flowing movement of the fingers through the sand, and the "added art of storyteller who keeps his audience in suspense, intriguing them with the arabesques and holding them breathless until the end of the story." To draw the *sona*, an array of dots is first constructed. Great care is taken to create equal distances between the dots by simultaneously using the index and ring fingers and then moving one of these fingers into the imprint left by the other. The continuous figure is drawn surrounding the dots without touching them. (In some very few cases, disconnected additional lines are drawn through some of the dots.) For several figures, the beginning and ending points of the Tshokwe tracings have been recorded but the exact tracing paths are known for very few. Before we examine some of the *sona*, we present aspects of the Tshokwe culture that help to comprehend the figures and their associated stories and names.

The Tshokwe are in the West Central Bantu culture area. Historically, culturally, and geographically they are closely related to the Lunda,

Lewena, Mbangala, Minungu, and Luchazi. Since the late nineteenth century, the Tshokwe have been the dominant majority in the Angoleis Lunda Sul/Lunda Norte regions (and just above that into Zaire) and spread more thinly in nearby areas (see Map 1). In 1960 this cluster of cultures contained about 800,000 people, of whom about 600,000 were Tshokwe.

The Tshokwe live in small villages under a family chief. The chiefdom passes to his younger brothers and then to his sisters' sons. His symbol of authority, which is also passed on, is a bracelet. The spirits of the ancestors (*mahamba*) and of nature are intermediaries between people and the Supreme Creator. The *mahamba* of the family of the community are represented by trees erected behind the chief's house and guarded by him. They are regularly given sacrifices and gifts of food. The individual families have smaller symbolic trees behind their own dwellings and small figurines are in cases or miniature huts around the village.

The *akishi* (sing. *mukishi*) are spirits that are incarnated in masks designated by the same name. When wearing a mask, the wearer is dedicated to and merged with the spirit. Many masks are associated with *mukanda*, a rite of passage of boys into adulthood. The rite begins when the chief of a village and his counselors decide there is a sufficiently large group of children. A camp, also called *mukanda*, is constructed in a specially cleared place. It is enclosed by a fence and contains round straw huts. After the ritual circumcision there are ritual dances, foods, clothes etc. The initiates then live in the camp for a year or two or even three, where they are taught, for example, rituals, history, and maskmaking. The mask Kalelwa, who had given the signal for the children to leave their homes, is in charge of the coming and going from the camp. The mothers are not allowed to see their sons and so Kalelwa makes loud noises to warn them away. The fathers can bring food and drink to the camp. After the prescribed education is complete, another celebration is held at which the young men are given new names, adult clothes, and then return home.

Several of the *sona* refer to the *mukanda*; three of these are shown in Figure 2.4. The simplest figure of all the *sona*, a continuous closed curve with no intersections, includes a story and exemplifies a Tshokwe topological concern. The figure (Figure 2.4c) depicts the camp. The line of dots are the children involved in the ceremony, the two higher dots the guardians of the camp, and the lower dots various neighbors or people not involved in the ceremony. The children and guardians are *inside* the camp; the others *outside*. One version emphasizes that the children cannot leave the camp, the other that the uninvolved people

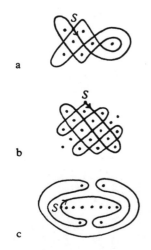

Figure 2.4. *Sona* related to the *mukanda*: a, the object of circumcision; b, subject unclear; c, the camp.

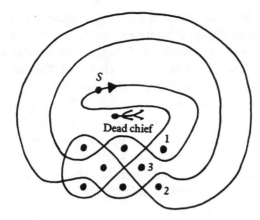

Figure 2.5. The succession of the dead chief. Dots 1, 2, and 3 are thieves. Only dot 3 can reach the village and steal the insignia.

cannot enter. A still lengthier version describes the fathers trying to bring food to the children while the *akishi* make ferocious noises in an attempt to frighten them away. No matter which version, most crucial to the story is that the figure, a simple closed planar curve, determines two regions of which it is the common boundary. That is what Western mathematicians call the *Jordan curve theorem*.

In other *sona* as well, the figure defines several regions in the plane and the story causes the audience to become aware that certain dots are interior or exterior to particular regions. The stories animate the dots in such a way that they become representative of any point within their region. For example, in the *lusona* shown in Figure 2.5, the chief of a village dies. Three thieves (dots labeled 1, 2, 3) try to steal the succession. The first two find that they cannot reach the chief but the third gets to the village, steals his insignias, and so succeeds him. Another *lusona* concerned with the separation of regions is actually a pair of *sona*. They attract our special attention because, in graph theoretic terms, they are *isomorphic*. A graph is defined by its vertices and interconnecting edges. It is the fact that two vertices are connected by an edge that is significant, not the length or curvature or color of the edge. Therefore, whether visually different or visually the same, two configurations with the same set of vertices similarly interconnected are graphically the same. To be more precise, two graphs are called *isomorphic* when their vertices and edges can be placed in one-to-one correspondence. Figure 2.6 shows a pair of graphs that differ visually but are isomorphic. The corresponding vertices have been labeled *A*, *B*, *C*, *D*, *E* and the same set of connecting

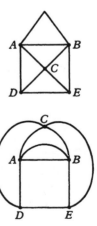

Figure 2.6. A pair of isomorphic graphs

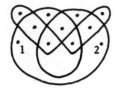

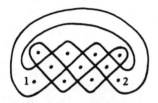

Figure 2.7. A pair of isomorphic *sona*. Dots 1 and 2 represent Sa Chituku and his wife, Na Chituku; the other dots are their neighbors. Sa Chitiku builds barriers that isolate his wife from the neighbors so that she will attend to cooking instead of visiting.

edges *AB, AB, AC, AD, BC, BE, CD, CE, DE* are found in both. The fact that the *sona* in Figure 2.7 are isomorphic is especially noteworthy because they share the same story and the same name. Thus, our mathematical notion of isomorphism is reflected in a Tshokwe conceptual linkage. A pervasive characteristic of the *sona* is the repetition of basic units variously combined into larger figures. The *muyombo* trees representing the village ancestors are the *sona* shown in Figure 2.8. The uppermost row of dots represents members of a family praying to the ancestors before undertaking some action; the rest of the dots are all the houses of the village showing that the entire life of the clan is enveloped by the influence of the ancestral spirits. The same *lusona* is found in

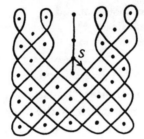

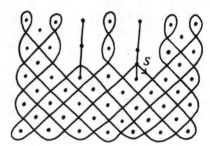

Figure 2.8. *Muyombo* trees

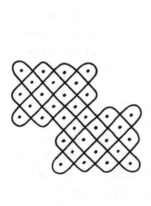

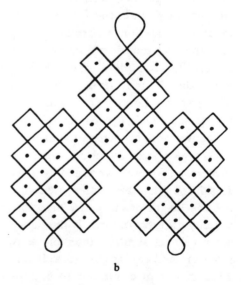

a   b

Figure 2.9. *Sona*: a, a hyrax in a rock; b, a leopard with his cubs

one-, two-, and three-tree versions. Comparing them, it is as if the smaller has been slit along its line of symmetry and another symmetric strip inserted. A different mode of extension is seen in a set of figures that has essentially one basic unit rotated and overlapped two, three, four, five, or six times. (Figure 2.9 shows those with two and three repetitions.) The *sona* in Figure 2.10 is unusually prominent and is even found in figures drawn on Tshokwe house walls. A set of *sona* in which four, six, eight, or nine of these units are variously connected are in Figure 2.11. The prominence of this *sona* is especially interesting because, without its dots, it is the very figure called *imbola* by the Bushoong children.

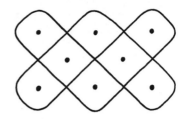

Figure 2.10. A small animal that lives in a tree hole and pierces the intestines

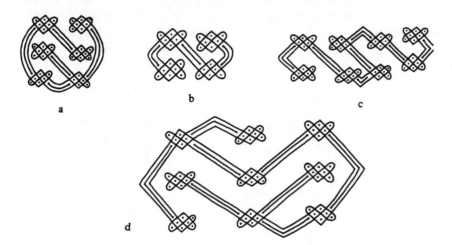

Figure 2.11. *Sona*: a, the *muyombo* tree of a former chief; b, the sanctuary of the *mahamba*; c, salt marshes; d, a labyrinth used in witchcraft

With few exceptions the Tshokwe *sona* are *regular* graphs of degree 4. (When all the vertices of a graph are of the same degree, the graph is *regular* and of that common degree.) Among the exceptions are some for which there are a pair of odd vertices. And, the fact that one of these corresponds to another of the three drawn by the Bushoong children, with the challenge that it be continuous with no retracing, provides further indication that there is some connection between the graphic concerns of the two groups. Three of the *sona* with a single pair of odd vertices are shown in Figure 2.12.

Ignoring the two flourishes, Figure 2.12c is the figure for which a general drawing procedure was stated earlier. The number of units along the first diagonal was designated by N; for the Bushoong figure N was equal to 10 while here *N* equals 8. Another Tshokwe figure is quite similar; the primary difference is that now *N* equals 6. Thus these constitute a set of three figures that share an overall configuration but differ due to the variation of a single parameter. Examples from three other sets are shown in Figure 2.13. Figure 2.13a is drawn on a 5 × 6 grid of dots; the other member of its set has the same configuration but is drawn on a 9 × 10 grid. The grid sizes for the set can be described as $(N - 1) \times N$, where *N* is 6 for the former and 10 for the latter. The *lusona* in Figure 2.13b is from another set of configurations drawn on grids of *N* rows of *N* dots each alternating with $N - 1$ rows of $N + 1$ dots each. Here *N* is 6, and for the other member of the set *N* is 7. Figure 2.13c comes from an even larger set of *sona*: these figures have rows of *N* dots alternating with rows of $N - 1$ dots for a total of $2N - 1$ rows. For Figure 2.13c, $N = 5$ and for other members of the set, $N = 2$, 6, and 8.

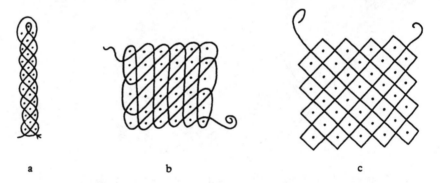

a   b   c

Figure 2.12. *Sona:* a, a large snake; b and c, intertwined rushes

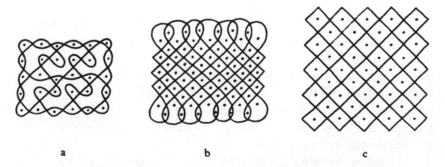

a   b   c

Figure 2.13. *Sona:* a, the marks on the ground left by a chicken when it is chased; b, fire; c, captured slaves surrounded during a night encampment

Some summarizing generalizations can be made based on the Tshokwe *sona* as a group.

1. Clearly the Tshokwe are interested in drawing figures continuously and in the use of curves to separate regions of the plane.

2. Dot placement, significant because it foreshadows the final figure, is an important part of the drawing procedure. Careful attention is given to the creation of rows and columns of equally spaced dots.

3. There are sets of *sona* that are similar in structure but varied in size. One type of set is characterized by a subfigure that is systematically repeated a different number of times in each *lusona* in the set; the other type is characterized by having a variable dimensional parameter.

4. The overriding feature of all of the Tshokwe *sona* is the prevalence of regular graphs of degree 4. Those few that have a pair of vertices of odd degree share another aspect: all other vertices in the figure are of degree 4. And what's more, the pair of odd vertices is *nonessential*; that is, a line can be drawn connecting the odd vertices without involving any other vertices or intersecting any edges of the figure. In short, they, too, are very close to being regular graphs of degree 4.

5 We now turn to another culture in another part of the world. Here, too, there is a tradition of tracing figures in the sand. The people are different; the figures are different; and the relationship of the figures to the culture is different. The concern for Eulerian paths is still central but the associated geometric and topological ideas are markedly different. What makes this case even more distinctive is that the exact tracing paths for the majority of figures are known to us. The ethnologist who first noted this figure tracing tradition, believing there was something most unusual in it, was meticulous in recording the specifics of almost 100 figures. This crucial information enables us to go beyond speculation or generalities and to see beneath the external features of completed figures.

The Republic of Vanuatu, called the New Hebrides before its independence in 1980, includes a chain of some eighty islands stretching over about 800 kilometers, with an indigenous population of close to 95,000. Malekula, one of the two largest islands, constitutes about one-fourth of Vanuatu's area. Europeans began to exploit this island about 1840. The social and psychological effects of colonial rule and the introduction of European diseases reduced the population by about 55 percent between the 1890s and 1930s. The culture area of interest here

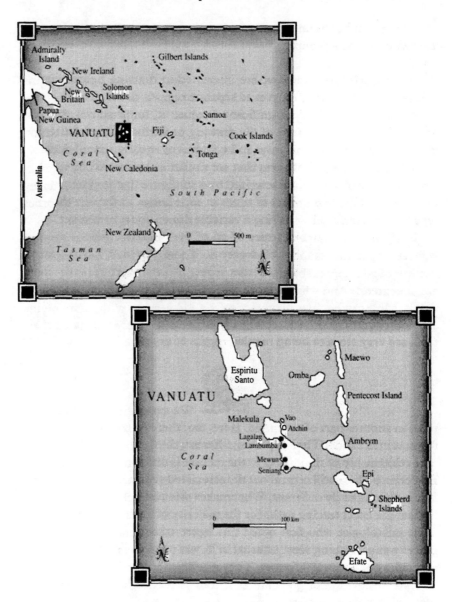

Map 2. Vanuatu in the South Pacific

includes principally Malekula and the islands of Vao and Atchin, just off its coast, but also extends to the nearby islands of Omba, Pentecost, and Ambrym (see Map 2). While there are some differences among them, all will be included under the Malekula.

The drawing of continuous figures is tightly enmeshed in the Malekulan ethos. Many of the figures are named for local flora and fauna but several are related to important myths and rituals. In addition to illustrating myths, there are even myths in which the drawing concept plays a significant role. The drawings, called *nitus*, are executed in the sand by men and knowledge of them is handed down from generation to generation. Often a framework of a few horizontal and vertical lines or rows of dots precedes the drawing but is not considered a part of the figure and, occasionally, lines for tails or such are added at the end. The figure is to be drawn with a single continuous line, the finger never stopping or being lifted from the ground, and no part covered twice. If possible, the drawing is to end at the point from which it began. When it does, a special word is given to it: the figure is *suon*.

The pervasive aspects of life in Malekula that appear in descriptions of the *nitus* are graded societies and pigs. The graded societies define a person's rank in the overall society. Men's societies have from ten to thirty ranks within them; women's societies are less elaborate with only about three to five ranks. Each rank confers certain privileges including designs and ornaments that one is allowed to wear and places one is allowed to sit. Each advance is achieved through the construction of monuments and wooden drums, various payments, sacrifices of tusked pigs, and the performance of ritual acts. Advancement is accompanied by public dancing and feasting ceremonies. Passage from rank to rank becomes progressively more difficult and indicates an increase in power, influence, fertility, and the aid of supernatural beings. The acquisition and breeding of tusked pigs is of great importance because they are needed in most ceremonies and rituals and they serve as payment for special services by sorcerers or craftsmen.

Among the Malekula, passage to the Land of the Dead is dependent on figures traced in the sand. Its exact place, its entrance, and who guards the entrance vary with the locale but generally the entrance is guarded by a ghost or spider-related ogre who is seated on a rock and challenges those trying to enter. There is a figure in the sand in front of the guardian and, as the ghost of the newly dead person approaches, the guardian erases half the figure. The challenge is to complete the figure which should have been learned during life, and failure results in being eaten.

Another myth describes the origin of death among humans. The myth centers around two brothers, Barkulkul and Marelul, who came to earth from the sky world. (They are two of five brothers collectively referred to as Ambat.) One day Barkulkul went on a trip but before doing so he took the precaution of enclosing his wife so that he would know of any intruder. In one version he uses a vine to make a spiderweb-like design on the closed door of the house; in another he loops a string figure around his wife's thighs. Marelul visits her during Barkulkul's absence and then improperly replaces the vine or string. Upon his return Barkulkul goes to the men's house and challenges all the men to draw figures in the ashes on the floor. Marelul's figure gives him away and Barkulkul kills him. Eventually Marelul returns to life but smells so badly from decay that others avoid him. He returns to the sky world where there is no death. Those that remain on earth are the ancestors of men who work for sustenance and eventually die. These tales provide the indigenous association of sand or ash tracings, string figures, vines, and spiderwebs. Some aspects of these tales are the subject of several figures. More significant, however, is that the tales emphasize the need to know one's figures *properly* and demonstrate their cultural importance by involving them in the most fundamental of questions—mortality and beyond death.

Viewed solely as graphs defined by vertices and edges, the *nitus* vary in complexity from simple closed curves to having more than one hundred vertices, some with degrees of 10 or 12. Of some ninety figures that

a   b

Figure 2.14.   a, a *nitus* associated with a secret society;
b, a ghost

I have analyzed, about ten are simple closed curves and another twenty are regular graphs of degree 4. About thirty more, while not regular graphs, have only even vertices, and then about fifteen each have a single pair of odd vertices. In addition, because tracing courses were documented, there is ample corroboration of the Malekula's stated concern for what we call Eulerian paths. For those figures with known tracing courses, where Eulerian paths are possible, almost all were traced that way. And, similarly, the stated intention of beginning and ending at the same point was almost always carried out where that could be done. A story associated with a *nitus*, in which retracing of some edges does occur, highlights the fact that the Malekula consider backtracking to be improper. The *nitus* is called "Rat eats breadfruit half remains." First a figure described as a breadfruit is drawn completely and *properly*. Then the retracing of some edges is described as a rat eating through the breadfruit. Using the retraced lines as a boundary, everything below it is erased as having been consumed. Thus, the retracing of lines carries forward a story about an already completed figure but has the effect of destroying parts of it.

Figures 2.14–2.16 show a few of the *nitus* that are simple closed curves or regular graphs of degree 4. The lattice-like figures have tracing procedures that are similar to those of the Bushoong; that is, they systematically proceed diagonally down and up as far as possible. Here, however, the changes in direction involve loops and curved segments and

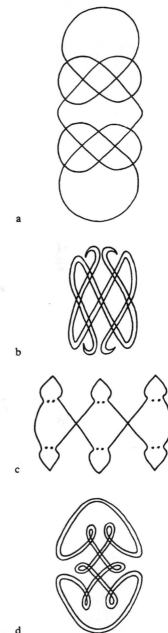

a

b

c

d

Figure 2.15. *Nitus*: a, the path that must be completed to get to the Land of the Dead; b, a flower headdress, first made by Ambat, worn in a funeral ritual; c, three sleeping ghosts; d, related to two secret societies and women's initiation rites

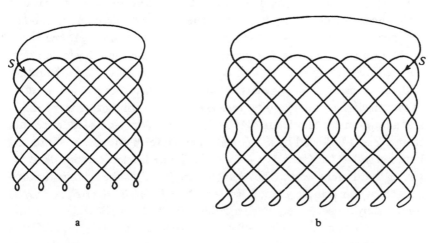

a

b

Figure 2.16. *Nitus*: a, untranslated; b, the nest of the hawk with eggs in the middle and tail feathers projecting from the bottom

there is a final sweep back to the starting point in order to end where one began.

6 Within individual figures the Malekula tracing procedures are systematic. But, far more important, there are larger, more general systems that underlie and unite groups of individual tracings. We will look at three of these extended systems, discussing each in terms of both the generalization that characterizes the system and some expressions of the generalization in particular *nitus*. The first two systems, taken together, account for the majority of *nitus* with a pair of odd vertices; the third involves figures that are more intricate in visual effect and in underlying concept. Keep in mind that tracing procedures are hand motions that leave transient marks in the sand. One way to talk about them or write about them is to refer to them by symbols. The symbols are ours, not theirs. But, as often happens in mathematics, if we assign our symbols with care, we can highlight logical and structural relationships that may otherwise escape us. It is the Malekula who created the tracing systems but it is we who are introducing symbols in an attempt to capture and convey the structure of their systems.

One set of six *nitus* is characterized by three basic drawing motions that are variously combined to give rise to a variety of figures. The *nitus* come from at least four islands and have diverse descriptions: a rat's track; a pig's head; a type of nut; the mark of the grandmother; octopus, food of the ghosts; and the stone on which is seated the ghost guarding the entrance to the Land of the Dead. Idealized versions of the three basic motions, denoted by the letters S, A, and B, are shown in Figure 2.17. Building with these basic notions, one circle-like unit results from S alone, a ladder of two circle-like units results from AB, a three-unit ladder from ASB, and a four-unit ladder from $AABB$. In general, an n-unit ladder results from $n/2$ A's followed by $n/2$ B's for n even; for n odd, an S is inserted between $(n-1)/2$ A's and $(n-1)/2$ B's. These ladders can be linked together in different ways. One *nitus* (Figure 2.18a) is the systematic connection of consecutively larger ladders (S, AB, ASB, $AABB$, $AASBB$) followed by consecutively smaller ladders ($AABB$, ASB, AB, S) resulting in a square array of circle-like units; another *nitus* (Figure 2.18b) is a rectangular array resulting from the systematic linkage of S, AB, AB, AB, AB, AB, AB, S. (In Figures 2.17 and 2.18, in order to identify the ladders on the *nitus*, each of the circle-like units is labeled i_j, meaning the

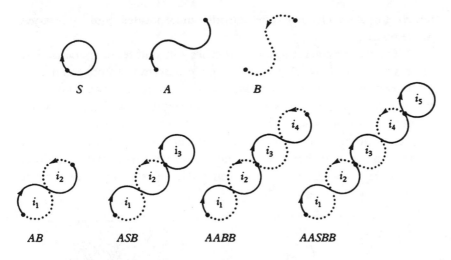

Figure 2.17. Basic motions *S*, *A*, and *B* combined into ladders of circle-like units. The subscripts denote the order in which the units are initiated.

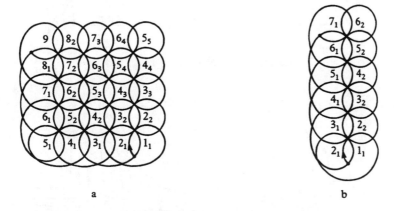

Figure 2.18. *Nitus*: a, the mark of the grandmother; b, a type of nut. Unit i_j is the *j*th unit initiated in the *i*th ladder.

*i*th ladder drawn in a specific *nitus* and the *j*th unit initiated in that ladder.) The other four *nitus* in the set are essentially the same except that they are built up from rotated versions of the three basic motions. And all of them differ slightly on their first and/or last few steps. Thus, in this system a few basic motions are formed into larger procedures, which are diversely yet systematically incorporated into still larger procedures.

The second tracing system underlies a second set of six *nitus*. The figures are from three islands and range in description from the nest of the burrowing ramé bird to a plaited coconut mat to a variety of yam. Each tracing begins with some initial procedure, that is, a particular sequence of drawing motions. Then, whatever the procedure, it is repeated within or around itself getting smaller and smaller or bigger and bigger. Figure 2.19 shows one of these *nitus* with its initial procedure. Another, less visually intricate because the repeated procedure never

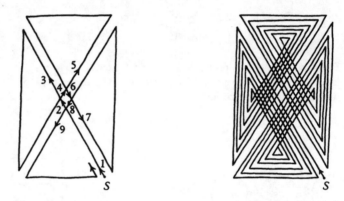

Figure 2.19. Nest of the *ramé* bird

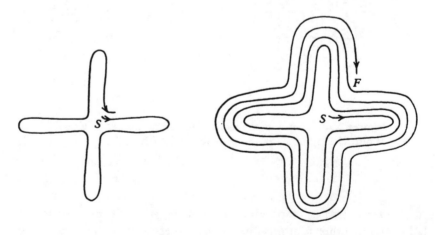

Figure 2.20. A type of yam

crosses a previously traced line, is shown in Figure 2.20. In the previous system, the same basic drawing procedures were combined in different ways; in this system it is the basic drawing procedures that differ while their manner of combination remains the same. The *nitus* in this group show the general tracing concept of an iterated procedure combined with systematic size modification. Also, the initial procedures, although different, share the feature of *almost* having bilateral or fourfold symmetry. In each, what interferes with the symmetry is the size modification just at the end of the procedure.

The third system is characterized by what I call a *process algebra*. The word *algebra* is used in its most fundamental sense: there are variable entities that are operated upon in accordance with specific rules. Here the variable entities are tracing procedures and the rules include processes that transform the procedures into other procedures. Furthermore, for the figures, only a distinct set of transformation processes is used. To give substance to this algebra, it is necessary to elaborate the specific set of processes involved. But first, since I am using them in a very specific way, the meaning of the words *procedures* and *processes* need to be made more explicit.

A tracing *procedure* is a particular sequence of motions used to draw a curve segment. It includes, therefore, both the directions of the motions and the order in which they occur. Figure 2.21a shows an example of a hypothetical tracing procedure; call it A. Procedure A is the sequence of motions: one unit up, one unit right, one unit diagonally up to the right.

A *process* is the way in which a tracing procedure is modified. If, beginning from the end of our hypothetical procedure A, the same procedure is repeated, the process is *identity* (no change), the resulting procedure is still A, and the overall procedure is A followed by A, which we symbolize by AA (see Figure 2.21b). If, however, when drawing the second segment every motion in A is rotated clockwise through 90° (up becomes right, right becomes down, diagonally up to the right becomes diagonally down to the right), the process is *90° rotation*, the new procedure is identified symbolically as A_{90}, and the overall procedure is AA_{90}. Another type of modification is the reflection of each motion across the vertical; that is, rights and lefts interchange while ups and downs remain the same . The result of a *vertical reflection* of A, which we denote by A_V, is shown in Figure 2.21d. You may have already noticed that these rotations and reflections look somewhat different from what we usually visualize when these words are used. That is because the procedure rather than the curve segment is being transformed, and the new procedure

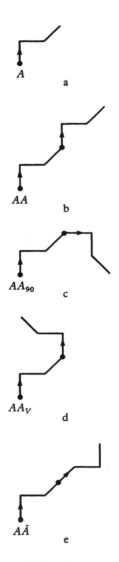

A a

AA b

AA_{90} c

AA_V d

$A\bar{A}$ e

Figure 2.21. A few of the processes

begins where the previous one ended. Another of the processes modifies the order of the motions rather than their directions. This process, here called *inversion*, reverses the order of the drawing procedure; whatever motion was last becomes first and so on until first comes last. (We use $\bar{A}$ for the inversion of procedure A and $A\bar{A}$ is shown in Figure 2.21e.) In all, the processes of this algebra are: no change or identity (I); reflection over a vertical line (V); reflection over a horizontal line (H); rotation through 90°, 180°, or 270°; and each of these can be, but need not be, simultaneous with inversion ($^-$). There are, therefore, twelve processes, as illustrated in Figure 2.22.

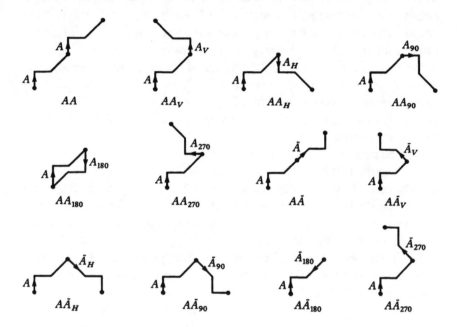

Figure 2.22. All of the processes

Now, turning to some individual figures in this group, each can be discussed in terms of the procedures and processes specific to it. For the *nitus* in Figure 2.23, when the initial procedure A (shown in the figure) is selected as the basic tracing unit, it becomes possible to concisely describe the overall tracing procedure as $AA_{90}A_{180}A_{270}$. Similarly, for the *nitus* in Figure 2.24, the overall procedure is $AA_{180}A_V A_H$ in terms of the initial procedure A we designated as its tracing unit.

Using the same basic tracing units, some alternative symbolic de-

A $AA_{90} A_{180} A_{270}$

Figure 2.23.   A yam

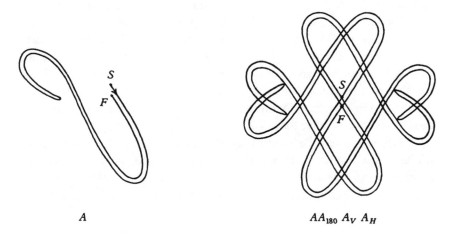

A $AA_{180} A_V A_H$

Figure 2.24.   Untitled

scriptions are posssible. In each of the foregoing descriptions, successive procedures were phrased in terms of the initial procedures. Each could be phrased, instead, in terms of the procedure that is its immediate predecessor. For the first *nitus*, that is:

$$AA_{90}(A_{90})_{90}(A_{180})_{90} \qquad \text{rather than } AA_{90}A_{180}A_{270}.$$

The new formulation rests upon the fact that a procedure modified by any process is itself a procedure that can, in turn, be modified. Thus the resulting procedure is the same whether one thinks of procedure A as having been rotated by 180° or thinks of procedure A_{90} as having been rotated by 90°. Another alternative is:

$$AA_{90}A_{180}A_{270} = (AA_{90})\underbrace{(AA_{90})_{180}}_{A_{180}A_{270}}.$$

This formulation reflects the following train of thought: procedure A is followed by procedure A_{90} and then their combined result is rotated by 180°. The formulation is possible because a procedure (in this case A) followed by another procedure (in this case A_{90}) is simply a new procedure (AA_{90}). These three symbolic formulations reflect three different conceptualizations but, nevertheless, all involve consecutive rotations. Taken together, the three versions highlight both the power and limitation of this algebraic approach. The power is that we can give expression to the structure of the overall tracing procedure. Further, symbolic manipulation enables different versions that reflect the same structure but different and equally plausible conceptualizations. The limitation, however, is that we have no way of selecting among different versions. In other words, the algebraic approach increases our understanding of what the Malekula actually did but does not uniquely translate into how they conceived of what they did.

Let us examine the description of Figure 2.24. For this, a little more symbolic manipulation is required. Again, the goal is to rephrase the description so that it is in terms of predecessor procedures rather than in terms of the initial procedure. That is, we will try to fill in the question marks for:

$$AA_{180}A_V A_H = AA_{180}\underbrace{(A_{180})_?}_{A_V}\underbrace{(A_V)_?}_{A_H} \quad \text{and in}$$

$$= (AA_{180})\underbrace{(AA_{180})_?}_{A_V A_H}.$$

We seek to find, for example, what process must act upon a 180° rotation to make the result appear the same as a vertical reflection. Table 2.1 is a summary of the results of all possible process pairings, and so within it we can find the answer. The column headings are the first process and the row labels are the second process.

Table 2.1

	I	90	180	270	V	H
I	I	90	180	270	V	H
90	90	180	270	I	(Y)	(X)
180	180	270	I	90	H	V
270	270	I	90	180	(X)	(Y)
V	V	(X)	H	(Y)	I	180
H	H	(Y)	V	(X)	180	I

Entries in the table were generated by (and can be verified by) actually carrying out the paired processes and then comparing the result with the individual processes. Thus, for example, a 180° rotation acting upon a 90° rotation appears the same as a 270° rotation, while a 180° rotation acting upon a vertical reflection appears the same as a horizontal reflection. (The circled entries X and Y are not among the individual processes and so the pairings leading to them could not have occurred. Had X and Y been present, they would have corresponded to reflections across diagonal lines.) The inversion process is not included in this table as it acts independently; that is, any procedure first inverted and then transformed by some process gives the same result as would inversion of the already transformed procedure. (For more details about inversion, see the notes to this section.)

Now we can fill in the question marks related to Figure 2.24. To resolve $(A_{180})_? = A_V$, we seek a V in the column headed 180° and then identify the row. Thus, we find $(A_{180})_H = A_V$. Similarly, for $(A_V)_? = A_H$, since the column headed V and row labeled 180° have H in their intersection, $(A_V)_{180} = A_H$. To solve $(AA_{180})_? = A_V A_H$, we first clarify a formality that has been implied but should be explicitly stated, namely,

$$(AB)_P = A_P B_P.$$

55

This means that for any of the processes (P) in the table, the result is the same whether procedures A and B are carried out sequentially and the combined result transformed, or whether each is transformed individually and then they are carried out sequentially. For our purposes then,

$$(AA_{180})_? = A_V A_H \qquad \text{can be rephrased as}$$
$$A_?(A_{180})_? = A_V A_H.$$

As we can guess and then verify in the table, this question mark should be replaced by a V. We have therefore found alternative equivalent descriptions focusing on initial procedure, preceding procedures, and preceding combination of procedures. They are:

$$AA_{180}A_V A_H = AA_{180}(A_{180})_H(A_V)_{180} = AA_{180}(AA_{180})_V.$$

Each provides a different conceptualization of the same overall procedure and, while the last seems more in keeping with some of the other tracings, choosing among the three would only be speculation.

Several elaborate figures are shown in Figures 2.25–2.28. Each is traced in three or four stages; that is, a continuous figure is drawn and then another, picking up from its endpoint, is drawn superimposed, and so on. Thus, each stage is a continuous figure in and of itself as is the total final figure. The interrelationship of the stages implies planning and a clear vision of the final goal. (In order to more fully savor the ideas of the Malekula, you are encouraged to try to trace some of these *nitus* without looking at the Malekula procedures. Then trace some of them using the Malekula procedures. Not only may you find their conceptions different from your own, but you may find many of their procedures quite graceful.) Figure 2.25 involves three stages, each a procedure and its 180° rotation. Its overall description is $AA_{180}BB_{180}CC_{180}$. (The first stage, viewed in finer detail, contains within it smaller procedures and their transforms: $A = X\bar{X}_V Y\bar{Y}_H$.) Another *nitus*, completed in four stages, is in Figure 2.26, described as $A\bar{A}B\bar{B}C\bar{C}D\bar{D}$. The processes of Figure 2.27 differ from stage to stage; this *nitus* has three stages and a small coda that is common to several of the figures. The stages are $(AA_V)(AA_V)_H$; $(BB_{270})(BB_{270}\text{truncated})_{180}$; and $C\bar{C}_V$. The emergence of such succint descriptions convinces us of the mathematical nature of the Malekula tracings. The procedures are both formal and logical but, what is more, they have structural simplicity while giving rise to complexity of visual effect.

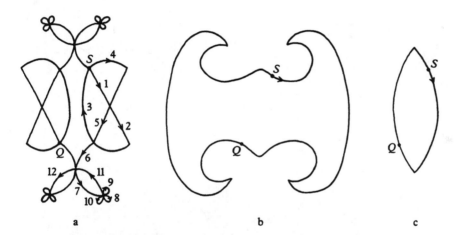

a   b   c

Figure 2.25. The *nitus* has three different names: "the stone of Ambat" and two others that refer to a story in which a mythical person attempts to kill and eat the Ambat brothers. Traced in three stages, each stage begins and ends at S. For each stage, the initial procedure goes from S to Q, and then the same procedure transformed by a 180° rotation goes from Q back to S. The complete procedure is $AA_{180}BB_{180}CC_{180}$, where $A = X\bar{X}_V Y\bar{Y}_H$ (X is segments 1–3 and Y is segments 6–11).

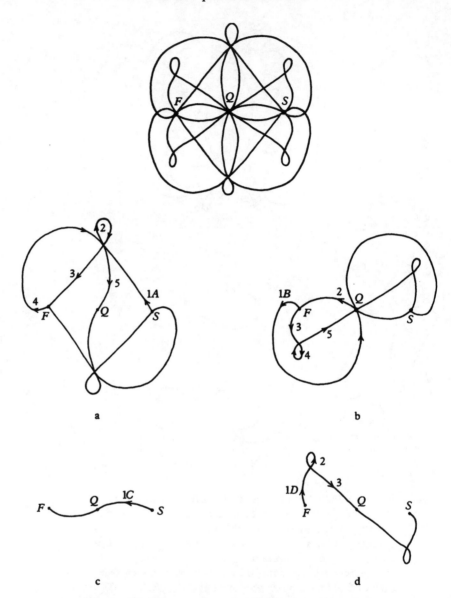

Figure 2.26. A turtle, traced in four stages. The first and third stages begin at S and end at F, and the second and fourth stages begin at F and end at S. For each stage, the initial procedure goes from S (or F) to Q, and then the same procedure transformed by inversion goes from Q to F (or S). The complete procedure is $A\bar{A}B\bar{B}C\bar{C}D\bar{D}$.

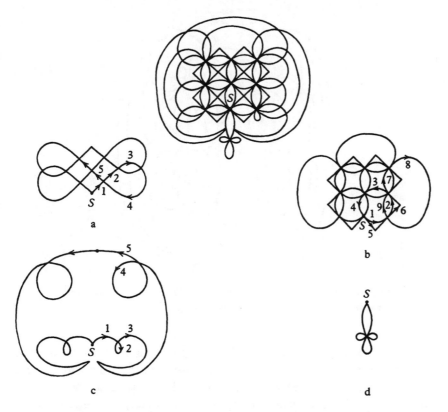

Figure 2.27. The banana stump, a tracing with three stages and a coda. Each stage begins and ends at S. The first stage is $(AA_V)(AA_V)_H$, where A is segments 1–5; the second stage is $(BB_{270})(BB_{270}$ truncated$)_{180}$, where B is segments 1–9; and the third stage is $C\bar{C}_V$, where C is segments 1–5.

This system encompasses a large number of the graphs with vertices of only even degree, including the very simple figures seen previously in Figure 2.14. The *nitus* in the group also have something else in common; with one exception, all are symmetric figures. Some have bilateral symmetry (a single axis of reflection) and most have double axis symmetry (a pair of perpendicular axes of reflection). Some of the latter also show 90° rotational symmetry and just one figure has 180° rotational symmetry without double axis symmetry. Here too, knowing the Malekula tracing procedures allows a deeper understanding. Rather than symmetry being an externally imposed concept based solely on *our* viewing of completed, static figures, symmetry can be seen as arising from

Figure 2.28.   Two fishes head to tail. The complete procedure
for this three-stage tracing is $AA_{180}BB_{180}CC_{180}$.

specific actions of the Malekula. This is especially significant because of
the interrelationship between how one traces a figure and what one sees
in it. Figure 2.23, for example, might be viewed as either double axis
symmetry or fourfold rotational symmetry. In fact, it resulted from a
tracing procedure involving rotation. Furthermore, we know (and are
possibly surprised to learn) that the unit of rotation was the curve segment from initial procedure A. In Figure 2.25, the visually ambiguous
case of double axis symmetry or 180° rotational symmetry was actually
brought about by 180° rotation. Looking at the apparent case of double
axis symmetry in the *nitus* of Figure 2.26, it is surprising to find that none
of its stages share that effect. Each of them shows 180° rotational symmetry brought about by the process of inversion! In the only *nitus* that
shows 180° rotational symmetry without double axis symmetry (Figure
2.28), the three-stage tracing procedure $AA_{180}BB_{180}CC_{180}$, the visual
effect, and the figure's description (a pair of like fishes head to tail) all
corroborate the 180° rotation.

 Some summarizing generaliza-
tions can be made based on the corpus of *nitus*.

1. Of the *nitus* for which it is possible to use a single continuous line and the courses are known, almost all were actually traced that way. This coincides with the stated intention of the Malekula. Another stated intention is, if possible, to end where one began. Of the *nitus* where that is possible and the courses are known, almost all do just that. And, backtracking was intentionally avoided, as underscored by erasure in the case where it occurs.

2. Our graph theoretic distinction between even-degreed graphs and those with a pair of odd vertices is correlated with a substantial division within the *nitus*. They are divided by the systems used to trace them as well as by the presence or lack of visual symmetry.

3. Symmetry is obviously of considerable importance as a large majority of the *nitus* show some form of it. (Those that do not are mainly among the graphs with a pair of odd vertices.) Furthermore, since the observed symmetries result from tracing procedures that involve reflections and rotations, we can conclude that visual symmetries were intentional on the part of the Malekula.

4. Within the Malekula's self-imposed constraints of symmetry, continuous paths, and ending at the starting point if possible, the individual figures are traced systematically. Further, for different groups of figures, these systematic tracings are particular expressions of larger systems. Each of the larger systems combines and/or transforms basic procedures into other procedures in ways that are both general and formal. One systematic mode of combination is used for the lattice-like figures; another is the systematic repetition of three basic procedures to form ladders with different numbers of circle-like units, which are then systematically formed into rectangular arrays. A general tracing scheme is the use of stages for the more complicated *nitus*, where the stages are essentially subgraphs with constraints similar to the graphs themselves. Over and above all of these, and used alone or in conjunction with them, are the transformations described as the set of processes including rotation, reflection, and inversion and the similitude transformation by which the sizes of the shapes are diminished or enlarged. The elements of this overall system are procedures to create shapes made up of curved and straight line segments; the transformations modify the direction or order in which the procedure is carried out or its scale. The procedures and transformed procedures are variedly yet systematically combined into larger procedures, that is, into paths that trace the corpus of the *nitus*.

5. In addition, related to or emerging from these formal aspects, there is a clear aesthetic component that we can appreciate.

8 Whether in the context of games, story telling, traveling to the Land of the Dead, abstract line systems, or Sunday strolls in Königsberg, different peoples have pondered the same problem. Each culture surrounded the idea of figures traced continuously with other geometric and topological ideas. Widely separated in space and in traditions, each of the three cultures, nevertheless, found the idea sufficiently intriguing to elaborate it well beyond practical necessity. This chapter, building around the shared underlying mathematical concept, has directed attention to the differences in culture, differences in context, and, hence, differences in elaboration.

Notes

2. For a brief introduction to graph theory, see O. Ore, *Graphs and Their Uses*, New Mathematical Library 10, Random House, N.Y., 1963, and for its history, see N. L. Biggs, E. K. Lloyd, R. J. Wilson, *Graph Theory 1736–1936*, Clarendon Press, Oxford, 1976. The idealized model of the Königsberg bridge problem shown in Figure 2.1d is usually attributed to Euler. However, according to R. J. Wilson ("An Eulerian trail through Königsberg," *J. of Graph Theory*, 10 (1986) 265–275), it first appeared in 1894 in *Mathematical Recreations and Problems* by W. W. Rouse Ball. Figure 2.1e appeared in T. Clausen, "De linearum tertii ordinis proprietatibus," *Astronomische Nachrichten* 21 (1844) cols. 209–216 and in J. B. Listing, *Vorstudien zur Topologie*, Göttinger Studien 1, 1847, 811–875. The former noted that a minimum of four lines was required and the latter stated the more general case that a figure with $2n$ odd vertices requires n lines. It was not until 1891, again using this figure as an example, that a proof for the general case appeared in E. Lucas, *Récreations Mathématiques*, Vol. 1, Gauthier-Villars et Fils, Paris. Figures 2.1a, b, and e are from a collection of Danish folk puzzles detailed in J. Kamp, *Danske Folkeminder, Aeventyr, Folksagen, Gaader, Rim Og Folketro, Samlede Fra Folkemende*, Neilsen, Odense, 1877. Particularly note the interplay of professional literature, mathematical recreation books, and Western folk culture. The statement by L. Wittgenstein is on p. 174e of his *Remarks on the Foundations of Mathematics*, G. H. von Wright, R. Rhees, G. E. M. Anscombe, eds., Blackwell, Oxford, 1956.

3. The Bushoong figures were collected by the ethnologist Emil Torday. They are reported in E. Torday and T. A. Joyce, *Notes Ethnographiques sur les Peuples Communément Appelés Bakuba, Ainsi que sur la Peuplades Apparentées les Bushongo*, Annales du Musée du Congo Belge, Ethnographie, Anthropologie, Serie 3, vol. 2, pt. 1, Bruxelles, 1910 and in E. Torday, *On the Trail of the Bushongo*, Lippincott, Philadelphia, 1925. The drawing processes for Figures 2.2b and 2.2c are not explicitly stated but can be reconstructed from the associated comments and figures in the former. My description of the Bushoong is drawn from Torday & Joyce (1910) and from M. J. Adams, "Where two dimensions meet: the Kuba of Zaire" in *Structure and Cognition in Art*, D. K. Washburn, ed., Cambridge University Press, N.Y., 1983, pp.40–55; Monni Adams, "Kuba embroidered cloth," *African Arts*, 12 (1978) 24–39; D. C. Rogers, *Royal Art of the Kuba*, University of Texas Press, Austin, 1979; J. Vansina, *Les Tribus Ba-Kuba et les Peuplades Apparentées*, Annales du Musée Royal du Congo Belge, Série in −8°, Tervuren, 1954; J. Vansina, *Le Royaume Kuba*, Annales du Musée Royal de L'Afrique Central, Série in −8°, Tervuren, 1964; and J. Vansina, "Kuba art and its cultural context," *African Forum*, 3–4 (1968) 13–17. The discussion in this section is similar to that in M. Ascher, "Graphs in cultures (II): a study in ethnomathematics," *Archive for the History of Exact Sciences*, 39 (1988) 75–95.

4. My description of the Tshokwe is drawn from M.-L. Bastin, *Art Decoratif Tshokwe I*, Publicacoes culturais no. 55, Museu do Dondo, Companhia de Diamantes de Angola, Lisbon, 1961; M.-L. Bastin, "Quelques oevres Tshokwe de musées et collections d'Allemagne et de Scandinavie," *Africa-Tervuren* 7 (1961) 101–105; A. Hauenstein, "La corbeille aux osselets divinatoires des Tchokwe (Angola)," *Anthropos* 56 (1961) 114–157; M. McCulloch, *The Southern Lunda and Related Peoples*, International African Institute, London, 1951. The *sona* have been collected and reported by a number of people. My analysis focused on those figures reported by more than one collector. My article, "Graphs in cultures (II)," cited above, contains detailed results of that analysis. See that article for additional details and for specific citations for the figures and associated stories used here. The *sona* in my study were from H. Baumann, *Lundabei Bauern und Jägern in Inner-Angola*, Würfel Verlag, Berlin, 1935; Th. H. Centner, *L'enfant Africain et ses Jeux dans le Cadre de la Vie Traditionelle au Katanga*, Collection mémoires CEPSI no. 17, Elizabethville, Katanga (Lubumbashi, Zaire), 1963; M. Fontinha, *Desenhos na Areia dos Quicos do Nordeste de Angola*, Estudos, ensaios e documentos no. 143, Instituto de Investigaçao Cientifica Tropical, Lisbon, 1983; E. Hamelberger, "Ecrit sur le sable," *Annales des Peres du Saint-Esprit*, 61 (1951) 123–127; G. Kubik, "Kulturelle und Sprachliche Feldforschungen in Nordwest-Zambia, 1971 und 1973," *Bulletin of the International Commission on Urgent Anthropological and Ethnological Research*, 17 (1975) 87–115; G. Kubik, "African graphic systems—a reassessment (Part II)," *Mitteilungen der Anthropologischen Gesellschaft in Wien (MAGW)*, 115 (1985) 77–101; G. Kubik, "*Tusona* ideographs—a lesson in 'objectivity' in interpretation," in *Archeolgical Objectivity in Interpretation 1*, World Archeological Congress 1986, Allen & Unwin, London, 1986, pp. 1–30; J. Redinha, *Peredes Pintadas da Lunda*, Publicações culturais no. 18, Museu do Dondo, Companhia de Diamantes de Angola, Lisbon, 1953; and E. dos Santos, "Contribuiçao para o estudo das pictografias e ideogramas dos Quicos" in *Estudos Sobre a Etnologia do Ultramar Português* 2, Estudos, ensaios e documentos no. 84, Junta de Investigações do

Ultramar, Lisbon, 1961. The quotation about the art of the story teller is from Centner (see citation above, p. 18).

5. In the 1920s, A. Bernard Deacon studied among the Malekula. With an eye and insight that were especially rare, he collected material that he believed demonstrated mathematical ability and evidence of abstract thought. One of the things he saw as mathematical was a complex drum signaling system based on identifying rhythms for each clan, rank, grade of pig, and certain special phrases, which were combined to uniquely identify each person and transmit entire messages understood by all of the adult males. Another was, in his words, "the amazingly intricate and ingenious" geometrical-figure drawings. He was meticulous in recording about ninety figures, including their exact tracing paths. Deacon died of blackwater fever as he was awaiting transportation home from his two years of field work. Fortunately a fellow graduate student, C. H. Wedgwood, and one of their professors, A. C. Haddon, published much of his work, including the annotated figures. The data for my analysis of the *nitus* were his annotated figures found in A. B. Deacon, "Geometrical drawings from Malekula and other islands of the New Hebrides," C. H. Wedgwood, ed., *Journal of the Royal Anthropological Institute*, 64 (1934) 129–175, as well as about twenty others reported in R. Firth, "A Raga tale," *Man*, 39 (1930) 58–60; A. C. Haddon, "The geometrical designs of Raga District, North Pentecost," pp. 143–147 in the Deacon article cited above; and J. W. Layard, *Stone Men of Malekula: Vao*, Chatto & Windus, London, 1942. The discussion here is drawn from my analysis which is reported in M. Ascher, "Graphs in culture: a study in ethnomathematics," *Historia Mathematica*, 15 (1988) 201–227. See that article for additonal details and for specific citations for the figures used here. Sources for the rest of the discussion were the work by Deacon cited above and A. B. Deacon, *Malekula, A Vanishing People in the New Hebrides*, C. H. Wedgwood, ed., Routledge, London, 1934; P. Gathercole, "Introduction," in M. Gardiner, *Footprints on Malekula: A Memoir of Bernard Deacon*, Salamander Press, 1984, pp. xi–xv; J. Guiart, "Société, rituels et mythes du Nord Ambrym," *Journal de la Société des Océanistes*, 7 (1951) 5–103; R. B. Lane, "The Melanesians of South Pentecost, New Hebrides" in *Gods, Ghosts and Men in Melanesia*, P. Lawrence and M. J. Meggitt, eds., Oxford University Press, 1965, pp. 250–279; H. Laracy, "Vanuatu" in *Historical Dictionary of Oceania*, R. D. Craig and F. P. Kings, eds., Greenwood Press, Westport, Connecticut, 1981, pp. 315–318; J. W. Layard, "Degree-taking rites in South West Bay, Malekula," *Journal of the Royal Anthropological Institute*, 58 (1928) 139–223; and P. E. Tattevin, "Mythes et légendes du sud de l'île Pentecôte (Nouvelles Hébrides)," *Anthropos*, 24 (1929) 983–1004. It is interesting to note that, according to Deacon, the term *nitus*, derived from the verb *tus* (to draw or paint), is so used by missionized Malekulans to refer to European writing. And, according to Guiart, string figures, which are made by adults, are designated by the same term as the sand drawings.

6. The procedures have been deemed variables of the process algebra because, within the general definition, they can be quite diverse. Further, to discuss the processes that apply to them, one need not know specifics of different procedures but only that they conform to the general definition. Symbolically stated, the relationship of inversion to the other processes is:

$$\overline{(A_P)} = (\bar{A})_P \quad \text{for} \quad P = I, 90, 180, 270, \text{V}, \text{H} \qquad \text{Also}$$

$$\overline{(\bar{A})} = A \quad \text{and} \quad \overline{(AB)} = \bar{B}\bar{A}.$$

c h a p t e r

The Logic of Kin Relations

t   h   r   e   e

Many ideas in mathematics are grounded in the concept of relations, that is, specified properties that link pairs of objects. For example, the relation *less than* holds for the pair of numbers 2, 3, the pair 1, 7, the pair 2, 9, and, in fact, for an infinity of pairs. The order in the pair is important; *less than* does not hold for the pairs 3, 2 or 7, 1 or 9, 2. Further, specific relations can be analyzed for characteristics they do or do not share with other relations. *Transitivity* is an example of one such characteristic. It means that the relation will satisfy the stipulation that if b is related to c in the specified way and if c is related to d in the same way, then b must be related to d in that way. *Less than* is a transitive relationship; if b is less than c and if c is less than

d, then, indeed, b is less than d. Being symmetric is another example of a characteristic of some relationships. For *symmetry* to be present, it must be true that if b is related to c in the specified way then c must be related to b in that same way. Clearly, *less than* is not symmetric; if b is less than c then c is not less than b. The concept of relations and their characteristics is sufficiently general to refer to many different types of relations among many different types of objects. It can easily refer to the most fundamental of human relations, the kinship of people.

The relationship *brother* is a good example. Is it a symmetric relationship? No. If John is Bill's brother then Bill is John's brother, but if John is Jane's brother, it does not follow that Jane is John's brother. We cannot, therefore, make the general statement that if b is the brother of c then c is the brother of b. This problem of different sexes does not occur with transitivity because the *brother* relationship is stipulated for b and c and then also for c and d. That is, b is the brother of c requires that b is a male, and c is the brother of d requires that c is a male. Then, whether d is male or female, it will follow that b is the brother of d. There is, however, another issue to be checked here and that is the question of different parents. If the relationship *brother* is confined to mean that both have the same pair of parents, then the relationship is transitive. If only one parent need be the same, the relationship is not transitive. Just examining these two characteristics of relations begins to make us think more about the meaning of these kinship terms and more about the kinship ties themselves. Before being introduced to a few more characteristics of relations, decide whether or not *cousin* is symmetric and/or transitive. The former yes, the latter no. What has become significant is that cousins can be from different *sides*, the father's side and the mother's side, and if from different sides they bear no relationship to each other.

The next idea about relations is forming *composites*; that is, looking at what results when coupling one relation with another. Both the composites *father of the father* and *father of the mother* are included in the relationship *grandfather*. As before, order is significant—*mother of the father* is not the same as *father of the mother*. Some composites are ambiguous; the father of my sister-in-law can be my father-in-law or can be the father of my brother's wife. And the result of the composite *brother of the brother* is just the same as *brother*.

2 Determining the results of combinations of relations is an exercise in logic that has even been cast as folk puzzles and used as entertainment. Here are just a few such puzzles from peoples with kin relations similar to ours.

- Two mothers and two daughters sleep in the same room. There are only three beds and each one sleeps on one of them. (Rio Grande do Sul, Brazil)
- One day three brothers were going past a graveyard. One of them said, "I shall go in so that I may say a prayer for the soul of my brother's son." The second man said the same thing. "I shall not go in," said the third brother. "My brother's son is not there." Who is buried in the graveyard? (Ireland)
- Who is the sister of my aunt, who is not my aunt, but is the daughter of my grandparents? (Puerto Rico)
- An old man was walking with a boy; the boy was asked, "How is the old man related to you?" He replied, "His mother is my mother's mother-in-law." What relation is that? (Vladimir Province, USSR)
- What relation to us is a brother-in-law to a brother of our mother? (Wales)

3 In all of these puzzles and all of the previous examples, the kin terms were known and familiar. More important is that an entire structure in which they are embedded was also known. It was from knowledge of the structure that decisions could be made about characteristics of the particular relations. Somehow, somewhere, as children, we incorporated that structure. But the structure contains a lot more than just the terms for kin and their logical relationships. Associated with them are different expectations, obligations, and modes of behavior. In general, one speaks in a different manner and about different subjects to a brother and a grandmother; one expects different things from a parent and a cousin; and you feel different obligations to a younger or older sibling and to your niece or godfather. From the kin structure there also emerge statements of whom you must not or ought not marry. There are laws regarding financial support, property ownership, inheritance, court testimony, and taxes that reflect and reinforce this structure. And religious rituals and ceremonies often have particular roles for those with certain relationships. For all of this, many other cultures have kin structures far more elaborate than ours. The structure may include more relations and extend to more people; it may be much more specific and much more diverse in terms of obligations, responsibilities, or modes of behavior; it may contain proscriptions of whom to marry as well as restrictions on whom not to marry; and it may have far more ceremonial, legal, economic, and political ramifications. For this reason, to understand the kinship system of a culture is to understand much about the culture. Here our focus is on just one aspect of these complex, multifaceted structures; our focus is on the logic of some relations. But even discussing just the logic will involve the introduction of some of the other facets.

4 The native peoples of Australia are renowned for their concern with kinship. One group that has a particularly complex kin system is the Warlpiri who live in a desert area in Australia's Northern Territory. The system encodes their social, political, and ritual organization and behavior. It encompasses everyone in the social environment and extends beyond that to their philosophical world.

The native Australians are hunters and foragers; they move about some defined territory in close coordination with the food and water supply. They travel in groups of from ten to sixty people but come together with related groups at various times for diverse occasions. Their personal possessions may be scant, but their spiritual and social worlds are rich and intricately ordered.

In native Australian cosmology, all that exists is part of an interconnected system. The system and pattern of life were set by the ancestors of the dreamtime who came from beneath the ground, from the sky, and from across the water. As they emerged and traveled across the continent, they formed mountains, rivers, trees, and rocks and named the plants and animals. The land boundaries of the tribes, the animals and plants that were to be sacred to each group, the sites that would be remembered in ceremonies and myths, all relate to the journey of the ancestors. The natural order of things, including the seasons of nature, the human life cycle, modes of human interaction, responsibilities for the land and for specific sites and ceremonies, all was specified by the ancestors. All of this, however, did not simply happen long ago. Through the dreaming, dreamtime continues and is renewed, reviewed, and reinforced. The dreaming is past, present, and future intertwined in a way too different from our own concepts to be clearly understood by us. By looking at the kin system, however, we may, perhaps, begin to understand.

5 The Warlpiri kin system has eight sections and each person is in one of them. Preferred marriages are to take place with a person from another specified section, and their children are in another section, which depends on the section of the mother. For now we will simply call the sections 1, ..., 8 and diagram the marriage rules as in Figure 3.1. A man in section 1, for example, marries a woman in section 5 and their children are in section 7. Or, a man in section 6 marries a woman in section 2 and their children are in section 3. Let us trace through several generations to see what happens. Since the outcomes vary depending on sex, we first follow several generations of women.

The mother of a section 1 woman will be in section 3, her mother will be in 2, her mother in 4, her mother in 1, and the cycle repeats. That is, the sections of consecutive generations of women form the matricycle

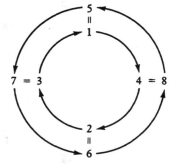

Figure 3.1. An equal sign indicates marriage partners. Each arrow points from the mother's section to the child's section.

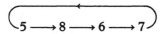

and so the same pattern is followed no matter where one starts within it. Similarly, there is another matricycle

$$5 \longrightarrow 8 \longrightarrow 6 \longrightarrow 7$$

Since the cycles do not overlap and each contains half of the eight sections, anthropologists refer to the group of four sections in each as a *moiety*. In this case, because of the mother relationship, each is a matrimoiety. The Warlpiri guidelines for interpersonal relations depend primarily on matrimoiety; interaction is relatively free with those in the same matrimoiety, while very elaborate conventions apply to interaction with those in the other matrimoiety. Figure 3.2, another diagram of the marriage rules, makes these matricycles and matrimoieties more explicit.

Now consider what happens for fathers and sons. A man from section 1 marries a woman from section 5 and his son is a 7; the son marries a 3 and his son is a 1. Thus, there is a father/son cycle of length 2:

$$1 \longrightarrow 7$$

Similarly, there are three other patricycles, each of length 2: (2, 8), (3, 6), and (4, 5). Inheritance rights and responsibilities for the land and religious rituals are determined by these groupings. Further, the set of four sections in the first two patricycles are joined into one patrimoiety and the sections of the latter two into another. Patrimoiety membership regulates activities in the politico-religious domain.

Figure 3.2. An equal sign indicates marriage partners. Each arrow points from the mother's section to the child's section.

Another Warlpiri grouping of the eight sections into two sets of four is {1, 6, 2, 5} and {8, 4, 7, 3}. These are generational moieties; those in each group are considered agemates. These moieties determine, for example, legal marriages and cooperative units for particular endeavors.

It is the matricycles and patricycles that are fundamental to the Warlpiri kin structure. They echo a fundamental tenet of native Australian cosmology; namely, they draw together the past and future into the present.

6 In order to delve more deeply into the implications and logical properties of this system, we introduce another descriptive mode. The previous chapter drew upon ideas from graph theory; here we draw upon ideas from group theory. Letter symbols will be used to represent the descent relationships and the eight sections. Just as the processes discussed in Chapter 2 transformed procedures into other procedures, here the descent relationships will link sections to other sections. As before, a table will be used to summarize the outcomes of consecutive relationships. This time, however, we will see that the Warlpiri kin system has a logical structure that Western mathematicians call a *dihedral group of order 8*.

To begin the labeling, we choose section 1 as the standard and call it I (identity, no change). The other sections will be viewed from this perspective, that is, through their relationships to I. Since the mothers of everyone in section I are in section 3, section 3 will be relabeled m for the relationship *mother*. Their mothers, who are in section 2, are the mothers of the mothers of the people in I, so section 2 becomes mm and then section 4 becomes mmm. Using f for father, section 7, with the fathers of those in I, is relabeled f and 5, 8, and 6 become mf, mmf, and $mmmf$ respectively. The order of the letters in a label is significant; mf, for example, is the section containing the mothers of the fathers of the people in I, and they are decidedly different from the fathers of their mothers. Finally, to make the labels more compact, a superscript denotes the number of consecutive times a letter appears; for example, m^3 is shorthand for mmm. Hence, what were

$$1, \quad 2, \quad 3, \quad 4, \quad 5, \quad 6, \quad 7, \quad 8 \qquad \text{are now}$$

$$I, \quad m^2, \quad m, \quad m^3, \quad mf, \quad m^3f, \quad f, \quad m^2f.$$

When looking for the section of the mothers of people in *mf*, we are using the relationship that leads from *I* to *m* (that is, mother) after having carried out those that led from *I* to *mf*. The result is just the same as had we gone from *I* to m^2f. Symbolically, this is:

$$m(mf) = m^2f.$$

In some other cases, the results are less straightforward; for example, the fathers of those in *mf* are in m^3, that is,

$$f(mf) = m^3.$$

The results of all possible pairings are in Table 3.1. (Each entry can be found or verified by following the diagrams in Figures 3.1 and 3.2.) The column heading indicates the first in a pair and the row label is the relationship applied to it. For the examples above, m^2f is in the intersection of the column headed *mf* and the row labeled *m*, while m^3 is in the column's intersection with the row labeled *f*.

Table 3.1

	I	m	m^2	m^3	f	mf	m^2f	m^3f
I	I	m	m^2	m^3	f	mf	m^2f	m^3f
m	m	m^2	m^3	I	mf	m^2f	m^3f	f
m^2	m^2	m^3	I	m	m^2f	m^3f	f	mf
m^3	m^3	I	m	m^2	m^3f	f	mf	m^2f
f	f	m^3f	m^2f	mf	I	m^3	m^2	m
mf	mf	f	m^3f	m^2f	m	I	m^3	m^2
m^2f	m^2f	mf	f	m^3f	m^2	m	I	m^3
m^3f	m^3f	m^2f	mf	f	m^3	m^2	m	I

A most important feature of these pairings is that *every* result is one of the original eight sections. (You may remember that the table of process pairs in Chapter 2 contained some *X*'s and *Y*'s that were extraneous to the processes.) The mathematical term for this important property is *closure*. Observe also that all the items in the column headed *I* replicate the row labels and, similarly, the row labeled *I* replicates the column headings. *I*, therefore, is deemed an *identity element*. Another important feature of the table is that *I* appears once in each row and each column. There is, therefore, a particular relationship (an *inverse*) linking each section back to the standard section *I*. For example, m^3 is the inverse of *m* (and vice versa) and m^3f is the inverse of itself.

Because this set of interrelated sections has the three aforementioned properties (closure, an identity element, and a unique inverse for each section) plus another called associativity, it constitutes what mathematicians call a *group*. *Associativity* means that consecutively applied relationships give the same result regardless of which is carried out first, as long as their order remains unchanged. For *mmf*, for example, the end result is the same section whether *m* is applied to *mf* or *mm* is applied to *f*; that is,

$$mmf = m(mf) = (mm)f = m^2f.$$

[Something familiar which is *not* associative is division with ordinary numbers. Carrying out $40 \div 4 \div 2$ as $(40 \div 4) \div 2$ gives a different result than does $40 \div (4 \div 2)$]. As a consequence of these defining properties, each element in the group occurs once and only once in every row and column in the table. Hence, among the Warlpiri, not only is each person in one of the eight sections but each section is linked to every section through some descent relationship. Since every pairing returns to one of the eight sections, all relationships resolve into eight relationships; no matter how extended the descent line, the results return and become merged with the closest of linkages. Three more properties of the kin system make it not just a group but the special group mathematicians call the *dihedral group of order 8*. The properties are:

$$m^4 = I; \quad f^2 = I; \quad \text{and} \quad (mf)(mf) = I.$$

The first is an expression of the matricycle; the second is the patricycle; and the last expresses the marriage rule. The preferred marriage is between people from 1 and 5, 2 and 6, 3 and 7, and 4 and 8, which were relabeled *I* and *mf*, m^2 and m^3f, *m* and *f*, and m^3 and m^2f. Notice that for each pair (say it is *X* and *Y*), if a *Y* is the spouse of an *X*, then

$$mf(X) = Y.$$

But *spouse* is a symmetric relationship, so it is also true that

$$mf(Y) = X.$$

And $(mf)(mf) = I$ expresses the fact that the spouse of one's spouse is oneself! That these properties are crucial to the logical structure of this system is underscored by the mathematical terminology: m and f are called the *generators* of this group and $m^4 = I, f^2 = I$, and $(mf)(mf) = I$ are its *defining relations*. Which section was originally labeled I makes no difference as it is these interrelationships that define the logical structure.

The name *dihedral group* comes from a geometric model with this same logical structure. A closely related example is the group whose elements are the $2n$ symmetries of a regular n-gon. In this case with eight elements, n is 4 and so the *regular* n-*gon* (an n-sided polygon with all angles equal) is a square. For the geometrically minded, the symmetries of the square are a useful spatial analogy to the Warlpiri kin system. If a square were rotated about its center through 90°, the result would be a square that looks quite the same; the placement of specific corners, however, would be different. Similarly, rotations through 180° and 270° would change the places of the corners but return the figure to the square. Reflection across a horizontal center line or vertical center line or either diagonal would also return the figure to the square. No matter how these rotations and reflections are mixed and matched, only the eight possibilities shown in Figure 3.3 can result. The corners are differently marked in the figure so that we can identify their positions after each motion. Designating some position as the standard I, calling a clockwise rotation through 90° m and a reflection across the horizontal line f, the other positions resolve into combinations of m and f. Basically there are three kinds of motions: rotations through 90° or multiples of 90°; reflections across lines perpendicular to the sides; and reflections across the diagonals. Four 90° rotations return to the original so $m^4 = I$; two reflections across a horizontal line return to the original so $f^2 = I$; and two reflec-

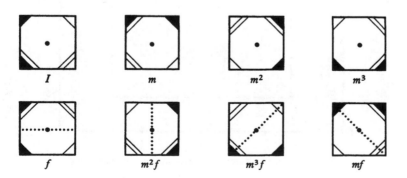

Figure 3.3. The symmetries of a square

tions across a diagonal line return to the original so $(mf)(mf) = I$. These are exactly the same generators and same defining relations that we saw before. The same table (Table 3.1) applies and the same group properties result. Although such different phenomena, the eight symmetries of the square and the eight sections of the Warlpiri kin system have the same logic of interrelationship.

Combining ideas from graph theory with ideas from group theory gives yet another way of visualizing a group; namely drawing a graph of the group. (The graphs are frequently referred to as *group diagrams* or *Cayley diagrams* after a nineteenth-century mathematician.) A graph, as discussed in the previous chapter, is a set of points (vertices) interconnected by lines (edges). A further distinction—specific directions of motion along the edges—is now added, making it a *directed graph*. Each element of the group is a vertex of the graph and the generators are the edges connecting them. For the Warlpiri kin system group, the directed graph is in Figure 3.4. Each section is a vertex of the graph. Moving along a solid edge in the direction of its arrow is the *mother* relation and, similarly, moving along a dotted edge in its indicated direction is the *father* relation. (Moving against the arrow would be the inverse of a relation.) Here the two matricycles and four patricycles become quite visible. To practice using this graph, find the sections of an I's father's mother's mother and her husband. (An I's father is in f, his mother is in mf, and her mother is in m^2f. The spouse of an m^2f is in $mf(m^2f)$, which is m^3.)

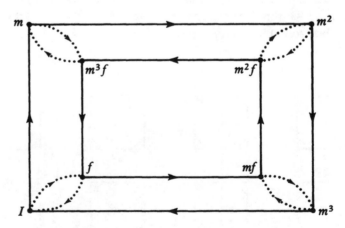

Figure 3.4. A graph of the group ($\rightarrow\!\!- m$, $\rightarrow\!\!- f$)

7 Whichever of the representations we use, we can answer a variety of questions. If we were in I, what are the sections of those we Westerners call our aunts and uncles? Some are in m and some are in f. Our mother's siblings are in m but their spouses are in f; our father's siblings are in f and so their spouses are in m. Some of our "first cousins," our mother's sisters' children and father's brothers' children, are with us in I, while the others, our father's sisters' children and mother's brothers' children, are in $m^3 f$. All of these cousins are in our generational moiety but then, too, so are our grandparents. (Our grandparents are in four different sections: m^2, mf, $m^3 f$, and I.) None of our parents, aunts and uncles, or first cousins are in the section with which we are to marry.

A Warlpiri, of course, does not go through this analysis. Each knows what absolute section he or she is in and learns what obligations and behaviors belong to being in that section. Also, each person knows behavior and obligations appropriate to different modes of relation and which of them goes with which section. Each person also knows, and that is the crucial point this discussion has tried to bring out, that everyone who is, was, or will be in the culture is bound to everyone else. Relatedness spreads beyond individuals to entire sections. The logic of the system cannot be expressed with our kin categories nor does it lend itself to our notions of close kin versus distant kin or your kin versus my kin. Obligations and behaviors are in the present but also extend backward and forward in time. They are obligations and behaviors to people but extend through them to the land, the flora and fauna, and whatever else, large or small, is in the environment. Every aspect of the culture—political, economic, religious—is encompassed by this system.

8 A variety of diagrams were used to describe the Warlpiri kin system. The system is theirs but the diagrams were ours. For the Malekula, whom we met in Chapter 2, we are fortunate to know of diagrams used by the people themselves to describe their kin system to an outsider. Their system is of particular interest here, as it is another that involves section membership, matricycles, and patricycles. In fact, the logical structure of the Malekula system is also a dihedral group, but of order 6 rather than 8.

At a time when kin systems of this type were little understood by Westerners, the elders of Malekula explained their system to an anthro-

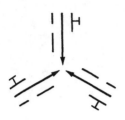

Figure 3.5. The unfinished Malekula diagram

Figure 3.6. The completed Malekula diagram

pologist using diagrams traced in the sand. One crucial aspect that they made clear was that terms such as *sister* referred not just to specific blood relatives but to others in the same section. In the explanation, an elder first drew three long lines arranged as equally spaced spokes of a wheel to represent men from each of three *bwelem*. The three men married, which he indicated by appending a short line to each of the long lines. Each couple had a boy and a girl who are of the same *bwelem* as the father but of the other "line," so longer lines, coming together at the center, were added and the children placed on the other side of them. His diagram, so far, is shown in Figure 3.5. The wives were from another *bwelem*. This he indicated by adding arrows to show where they came from. But, the elder emphasized, marriage moves in both directions. The males on the diagram who were yet without wives would, therefore, marry the sisters of the men who had married their sisters. This he showed by adding three short lines and arrows from them. The final diagram is Figure 3.6.

Following the association of wives, mothers, and daughters in the outer circle, we see a matricycle of length 3. Another matricycle is in the inner circle. The two cycles do not overlap and so there are six sections forming two matrimorieties ("lines"). A *bwelem* is a father-son pair, with each in a different "line" and forming a cycle of length two. If we start anywhere and label someone's section I and his/her father's section f, since I and f are in different "lines," the two matricycles are:

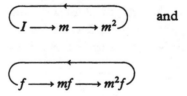

and

Following the logic of the diagram, the patricycles are (I, f), $(m, m^2 f)$, and (m, mf) and marriages take place between X and Y such that

$$mf(X) = Y; \quad \text{and} \quad mf(Y) = X.$$

Thus, we once again have a group but this time it has six elements $(I, m, m^2, f, mf, m^2 f)$ and the pairings shown in Table 3.2.

The Malekula diagram is repeated with section labels in Figure 3.7 to make more apparent the correspondence between our symbols and theirs. Their diagram is a clear and concise solution to an essentially mathematical problem; namely, it transforms a conceptual model into a planar representation. Comparing it with our directed graph of the group underlying the kin system (Figure 3.8), we see some similarity. Theirs, however, also includes marriages and male/female distinctions.

Not only is the logical structure of the system a group, it is a dihedral group of order 6. The generators are m and f and the defining relations are

$$m^3 = I; \quad f^2 = I; \quad \text{and} \quad (mf)(mf) = I.$$

These generators and three relations are the same as those we saw before but with one exception. Here $m^3 = I$ instead of $m^4 = I$. In general, for a

Table 3.2

	I	m	m^2	f	mf	m^2f
I	I	m	m^2	f	mf	m^2f
m	m	m^2	I	mf	m^2f	f
m^2	m^2	I	m	m^2f	f	mf
f	f	m^2f	mf	I	m^2	m
mf	mf	f	m^2f	m	I	m^2
m^2f	m^2f	mf	f	m^2	m	I

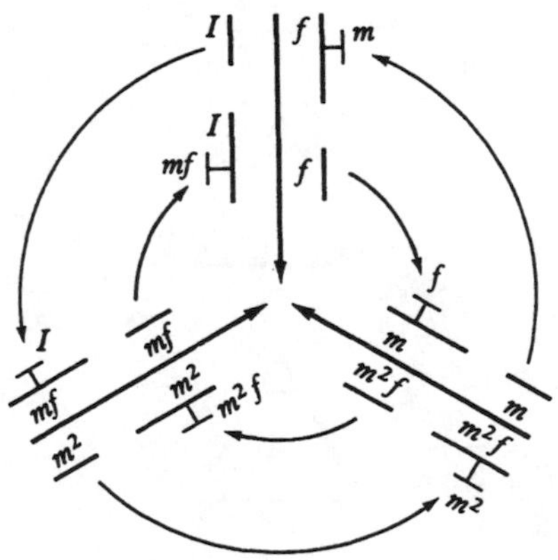

Figure 3.7. The Malekula diagram with sections labeled

dihedral group, the defining relations are

$$m^n = I; \quad f^2 = I; \quad \text{and} \quad (mf)(mf) = I$$

and the order of the group is then $2n$. So, both the kin structures are dihedral groups but of different orders. In this case, for $n = 3$, the spatial analogy of symmetries of a regular n-gon is the symmetries of an equilateral triangle. Rotations through 120° and 240° and reflections across the medians are the basic symmetry transformations of the triangle (see Figure 3.9).

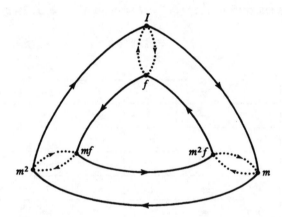

Figure 3.8. A graph of the group $(\rightarrow\!\!\!-\, m, \rightarrow\!\!-f\,)$

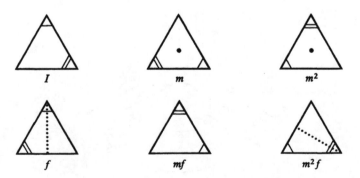

Figure 3.9. The symmetries of a triangle

9 This discussion of the logic of kin relations began with relationships between individuals and with familiar relationships from our own culture. Kin systems, quite different from our own, from two other cultures, draw upon the mathematical concept of a dihedral group. Most important is that the members of each culture share a conceptual model and draw inferences from it. Among the Warlpiri, the structure and all that such a structure implies has ramifications throughout the culture. For the Malekula, we know at least that the sand-tracing tradition associated with mythology reiterates some of the ideas implicit in the kin structure—ideas of formal procedures, symmetry, and continuous lines ending where they began.

To me it is striking to find that a logical structure studied abstractly and extensively by Western mathematicians plays a central and significant role in the day-to-day life of some peoples. And, what is possibly more surprising is that the realm of the logical structure is human relations raher than the physical or technological.

Notes

2. The Brazilian puzzle is no. 1.304 in Fausto Teixeira, *O Livro Das Adivinhas Brasileiras*, Editôra Letras e Artes Ltda, Rio de Janeiro, 1964. The same puzzle, but with fathers and sons, is found in Ilhan Basgöz and Andreas Tietze, *Bilmece: A Corpus of Turkish Riddles*, University of California Press, Berkeley, 1973, no. 1043. The Irish puzzle is no. 631a in Vernam Hull and Archer Taylor, *A Collection of Irish Riddles*, University of California Press, Berkeley, 1955. The Puerto Rican puzzle is no. 344 in R. R. de Arellano, *Folklore Portorriqueño*, Junta Para Ampliación de Estudios e Investigaciones Científicas, Centro de Estudios Históricos, Madrid, 1926. No. 333 in Vernam E. Hull and Archer Taylor, *A Collection of Welsh Riddles*, University of California Press, Berkeley, 1942 asks about the sister of an uncle instead. The Russian puzzle is no. 2364 in *Riddles of the Russian People* collected by D. Sadovnikov and originally published in 1876, English translation Ann C. Bigelow, Ardis Publishers, Ann Arbor, 1986. The last puzzle is no. 331 in the Welsh collection cited above.

4. For further discussions of native Australian philosophy and world view see A. P. Elkin, "Elements of Australian aboriginal philosophy," *Oceania* 40 (1969) 85–98; Jennifer Isaacs, compiler-editor, *Australian Dreaming*, Lansdowne Press, Sydney 1980; and W. E. H. Stanner, "The dreaming" in *Australian Signpost*, T. A. G. Hungerford, ed., F. W. Cheshire, Melbourne, 1956, pp. 51–65.

5. My discussion of the kin system of the Warlpiri is based particularly on the information in Mary Laughren's "Warlpiri kinship structure" in *Lan-*

guages of Kinship in Aboriginal Australia, Jeffrey Heath, Francesca Merlan, and Alan Rumsey, eds., Oceania Linguistic Monographs, No. 24, University of Sydney, 1982. Other useful items in the same monograph were the "Introduction" by J. Heath, pp.1–18 and F. Merlan's " 'Egocentric' and 'altercentric' usage of kin terms in Maṇarayi," pp. 125–140. For further details on native Australian kinship this monograph is recommended. My appreciation to Barry Alpher who first described to me in 1979 the Aranda descent and marriage rules and later directed me to the above references. Additional references used were T. G. H. Strehlow, *Aranda Traditions*, Melbourne University Press, Melbourne, 1947, and Jean Kirton, with Nero Timothy, "Some thoughts on Yanyuwa languages and culture" in *Language and Culture*, S. Hargrave, ed., Work Papers of the Summer Institute of Linguistics-Australian Aborigines Branch, series B, vol. 8, Darwin, 1982, pp. 1–18.

6. For a basic introduction to groups, graphs of groups, and dihedral groups see Israel Grossman and Wilhelm Magnus, *Graphs and Their Groups*, The New Mathematical Library, Mathematical Association of America, Washington D.C., original edition Random House, 1964, or Chapter 1 in Wilhelm Magnus, Abraham Karrass, and Donald Solitar, *Combinatorial Group Theory*, 2nd revised edition, Dover Publications, Inc., N.Y., 1976. For a more advanced discussion see Chapters 1 and 2 and pp. 115–117 in John S. Rose, *A Course on Group Theory*, Cambridge University Press, Cambridge, 1978.

The identification of the Warlpiri kin system as a dihedral group of order 8 is from the article by Laughren (cited above). She also involves the concepts of cosets, subgroups, and factor groups. A basic work on mathematical models of kinship is P. Courrège, "Un modèle mathématique des structure élémentaires de parenté" in *Anthropologie et Calcul*, P. Richard & R. Jaulin, eds., Union Générale d'Éditions, Paris, 1971, pp. 126–181. One of the examples in it is the kin system of the Aranda, a group closely related to the Warlpiri.

8. The Malekula diagrams and explanations were reported by A. Bernard Deacon who so carefully recorded the sand tracings. In a letter to A. C. Haddon he wrote: "the older men explained the system to me perfectly lucidly, I could not explain it to anyone better myself.... The way they could reason about relationships from their diagrams was absolutely on a par with a good scientific exposition in a lecture room." This quotation is from the "Preface" written by A. C. Haddon in A. Bernard Deacon, *Malekula, a Vanishing People in the New Hebrides*, edited by C. H. Wedgewood, George Routledge & Sons, London, 1934, pp. xxii–xxiii. The diagrams and discussion of the kinship system are in A. B. Deacon, "The regulation of marriage in Ambrym," *Journal of the Royal Anthropological Institute*, 57 (27) 325–342. The system is described as a dihedral group of order six in Georges Guilbaud, "Système parental et matrimonial au Nord Ambrym," *Journal de la Société des Océanistes*, 26 (1970) 9–32. An overview of the writings about the kin system is in Paul Jorion, "Alternative approaches to the Ambrymese kinship terminology" in *New Trends in Mathematical Anthropology*, Gisèle de Meur, editor, Routledge & Kegan Paul, London, 1986, pp. 167–197.

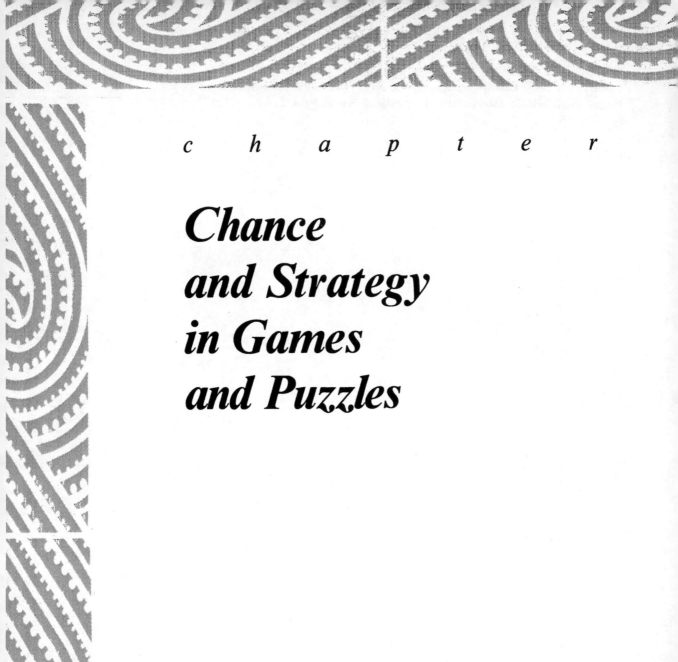

c h a p t e r

Chance and Strategy in Games and Puzzles

f   o   u   r

1 The mention of games brings
to mind a large and varied set of activities—children's street play,
puzzles, board games, dice games, card games, word games, golf, team
sports, and international competitions. In general, the activities we call
games have clearly defined goals toward which the players move while
following agreed upon rules. We can further classify games into those
involving physical skill, strategy, chance, or combinations of these. Since
our interest is mathematical ideas, we exclude those games solely involving physical skill and those depending on information other than the
rules of play. The games, then, that we deem somehow mathematical are
those that depend on chance or those in which the strategies must rely

on logic. In our culture, some popular games that rely solely on chance are matching coins, craps, bingo, and roulette. In these the players may bet but they make no choices that affect the outcome of the game. Whether they win or lose is beyond their control. In the games of strategy such as tic-tac-toe, checkers, or chess, on the other hand, each player does make choices; the choices rely on the logical implications of the moves in terms of progress toward the end goal and opening or closing further choices for oneself or one's opponent. For games that combine chance and strategy, such as bridge or poker, the evaluation of moves must also take into consideration the possible effects of certain chance factors.

To any game, however, there is more than just a formal statement of goals and rules. Consider in their entirety the games just cited from our culture. There are specific yet simple objects needed to play. But even the simple objects are sometimes quite elaborate and may become

Figure 4.1. A chessboard at the Mathematisches Forschungs-Institut Oberwolfach. The pieces are about three feet high.

personally prized possessions. There are times and places that the play of each game is appropriate and other occasions when it is viewed as outrageous. Although played by pairs or groups of people, onlookers are frequently involved. Each of the games is usually associated with a particular social setting, which may even be typified by different foods, different smoking habits, or different haircuts—a bridge party, a crap game, and a chess match bring quite different images to mind. Each of the games can also be played with different levels of concentration; they can be elaborated into regional tournaments; or they can be surrounded by auxiliary rewards that are not specified in the rules.

It is all such aspects taken together that make up a game. In all these ramifications, each game can be seen to be an expression of the culture. But the games are tied to their culture in a deeper way. In one cross-cultural study, games of strategy have been viewed as models of social interaction. A chessboard, for example, may be thought of as a field of battle on which two armies fight in defense of their royal leaders. The mobility and value of each piece depend on its social status. Another example, the Mancala game, popular throughout Africa, is sometimes used to demonstrate a chieftain's strategic abilities or even to decide who will be chief. Quite different from chess, however, the Mancala onlookers (as well as the players) are expected to contribute to a noisy, distracting atmosphere. In contrast to games of strategy, games of chance have been deemed models of interaction with the supernatural and are often linked with religion. Just ask people today, in our culture, to what they attribute their good or bad luck of winning or losing games of chance, or why some think such games are sinful. Each culture, then, creates different games and embeds them differently. Nevertheless, games of strategy are evidence of the enjoyment of logical play and logical challenge, and in games of chance there is implicit involvement with concepts of probability.

Now we will look in detail at a simple but widespread Native American game of chance and then at a Maori game of strategy. To close the chapter we will discuss a ubiquitous logical puzzle found in Western mathematics recreation books, in Western folk culture, and in several variants in African culture.

2 A dish and some small flat disks are the objects used in a game of chance widespread among Native Americans. The name of the game varies but usually reflects the type of objects used: peach stones among the Cayuga; deer buttons made of rounded and polished pieces of elk horn among the Seneca; shaped pieces of mussel shell among the Hupa; and butter beans among the Cherokee. Whatever the objects used, they are shaped and decorated or colored so that each has two faces that are distinguishable from each other. The number of disks also varies from group to group but usually there are six or eight.

There are two players. One of them places the disks in the dish and, by striking or shaking the dish, causes the disks to jump and resettle. The resulting assortment determines the number of points won and whether or not the player goes again or must pass the dish to his opponent. Just as the number of disks varies, so do the point values assigned. Quite often, an auxiliary set of sticks or beans serves as counters to keep track of the points won. We will concentrate on the version of this game found among some Iroquoian groups in what are now the northeastern United States and Ontario Province, Canada. Following some early European writings, we call it the game of Dish.

Among the Cayuga, the original inhabitants of the area in which I now live, the dish was a wooden bowl and the disks were six smoothed and flattened peach stones blackened by burning on one side. The auxiliary counters were beans. If the tossed peach stones landed with all six faces showing the same color (six black or six neutral), the player scored five points. For five faces of the same color (five black and one neutral or five neutral and one black), the player scored one point. In each of these cases, the player also earned another toss. For all other results, the player scored no points and had to pass the bowl to the opponent. Some prearranged total number of points, ranging from 40 to 100, determined the winner of the game.

Before describing the context of the game, let us look at some of the probabilistic implications of these scoring rules. When discussing probability, the focus is first on all possible outcomes and then on what fraction of them are of a particular type. For just one peach stone with two faces, there are only two possible outcomes: the black face (B) or the neutral face (N). Assuming no bias so that each face has the same chance of showing up, each has a probability of 1/2. Then, for two peach stones, there are four equally likely outcomes: both black (BB); a black followed by a neutral (BN); a neutral followed by a black (NB); or two neutral (NN). Two blacks are one possibility out of these four so its probability is 1/4; two neutral are one out of the four so its probability is 1/4. One of each color, however, is two out of the four (BN or NB) and so for that the probability is 2/4. Alternatively, we could arrive at the same measures of probability by saying that just one peach stone is B or N, each with the probability of 1/2, and the second is again B or N, each with a probability of 1/2 so: *B* followed by *B* has probability 1/2 *of* 1/2 or 1/4; *N* followed by *N* is 1/2 *of* 1/2 or 1/4; and, since one of each color can be arrived at two different ways, its probability is 2 *times* (1/2 *of* 1/2) or 2/4. For three peach stones there can be *B* or *N* followed by *B* or *N* followed by *B* or *N* or, in all, the eight outcomes

BBB, BBN, BNB, BNN, NBB, NBN, NNB, NNN.

Hence, getting three *B*'s has probability 1/8, getting two *B*'s and one *N* has probability 3/8, getting one *B* and two *N*'s has probability 3/8, and getting three *N*'s has probability 1/8. Continuing this line of reasoning, there are 64 possible outcomes when six peach stones are tossed. The outcomes and their probabilities are:

Table 4.1

Outcomes	Probability
6 *B*'s	1/64
5 *B*'s and 1 *N*	6/64
4 *B*'s and 2 *N*'s	15/64
3 *B*'s and 3 *N*'s	20/64
2 *B*'s and 4 *N*'s	15/64
1 *B* and 5 *N*'s	6/64
6 *N*'s	1/64

In a general way, the point values assigned by the Cayuga reflect these probabilities; the least likely outcomes earn the most points and the most likely earn the least points. What is striking, however, is the closeness of the point values to the specific values that would result from a probabilistic assessment. If we use as a baseline the assignment of 5 points for getting all blacks, then getting all neutrals should and does receive 5 points. Getting five blacks should receive 1/6 as many points, as it is six times as likely to occur. That would mean it should receive 5/6 points and, in fact, among the Cayuga, it gets the closest integer to that, namely one point. Again, since five neutrals have the same probability as five blacks, they receive the same score. The occurrence of four blacks (or neutrals) is fifteen times as likely as all blacks (or neutrals) and so should get 1/15 of the 5 points or 5/15 points. Here the closest integer is 0, and that is just what the Cayuga assign to it. Finally, getting three blacks is twenty times as likely as getting all blacks and so should get 1/20 of the 5 points or 1/4 point, and so, rounded to the closest integer, it too receives 0 points. In all, the comparison of the probabilistically derived and actual point values is:

Number of B's (or N's)	6	5	4	3	2	1	0
Probabilistic point value	5	5/6	1/3	1/4	1/3	5/6	5
Actual point value	5	1	0	0	0	1	5

The close correspondence of the point values used by the Cayuga and those based on our ideas of probability strongly implies a probabilistic basis for the Cayuga's choices. Were they based on some other unrelated criteria, we would not expect such a close match.

We can also use our ideas of probability to get a sense of the pace of the game. In all, six of a kind or five of a kind win some points and lead to another toss by the same player. From our table of probabilities (Table 4.1), we can see that the chances are 14/64, or just over 20 percent, that a player will gain something on a single toss. Winning tosses, therefore, occur often enough, but not too often, to make for an interesting game.

Another probabilistic concept, *expected value*, can give us an idea of the point worth of each toss and so some idea of the length of the game. The *expected value* of a toss is the weighted average of all possible results where the probabilities of the different results are the weights used for them. For example, if on the toss of a single die, you were given $3 if a 1, 2, 3, or 4 appeared and nothing for a 5 or 6, the expected value of a toss would be

$$4/6(\$3) + 2/6(\$0) = \$2.$$

There would be no result for which you actually would be given $2, but the *weighted* average of the $3 and $0 is $2. In the game of Dish, on a single toss, a player gets 5 points with probability 2/64, 1 point with probability 12/64, and no points with probability 50/64. The expected number of points is

$$5(2/64) + 1(12/64) + 0(50/64) = 24/64$$

and so a total score of about 40 points would be expected to take many, but not too many, tosses.

Clearly, the fact that certain results enable a player to throw again adds variety and interest to the game. With the help of a tree diagram, we can trace through what can happen before a player must surrender the dish. Figure 4.2 shows three successive tosses. On the first, the resulting points are 0, 1, or 5. With 1 or 5, the player continues to a second toss and, whichever it was, again a 0, 1, or 5 can result. Another 1 or 5 enables the player to continue to a third toss. The possible number of points amassed in three successive tosses can be found by following down the various branches of the tree. Fifteen points result from scoring 5 followed by 5 followed by 5; 11 points from scoring 5 followed by 5 followed by 1 (or 1, 5, 5, or 5, 1, 5); and so on. In all, the three tosses could end the player's turn with 2, 6, or 10 points or yield 15, 11, 7, or 3 points plus yet another toss.

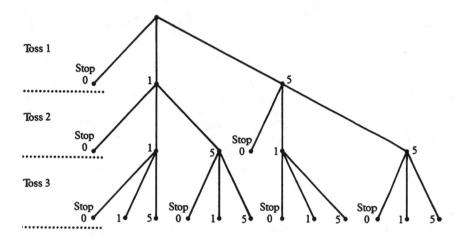

Figure 4.2. Three consecutive tosses

Figure 4.3 again shows three successive tosses. This time, however, the tree includes the probabilities along with the individual point value outcomes. From these, the composite probabilities of all possible results of one, two, or three consecutive tosses can be calculated. We use the same basic principles we used earlier, namely:

1. The probability of a specific string of consecutive independent outcomes is the product of their individual probabilities.
2. The probabilities are summed for the various specific strings that result in the same total point value.

The probability, for example, of getting 0 points and stopping on the first toss is 50/64 (or 78.13 percent) whereas getting 1 point on the first toss and then 0 and stopping on the second is 12/64 *times* 50/64 (or 14.65 percent). Two different routes would lead to stopping on the third toss with 6 points; it could be 1 point followed by 5 points followed by 0 or

5 points followed by 1 point followed by 0. Its probability, therefore, is

$$(12/64)(2/64)(50/64) + (2/64)(12/64)(50/64) = 75/8192 \text{ (or .92 percent)}.$$

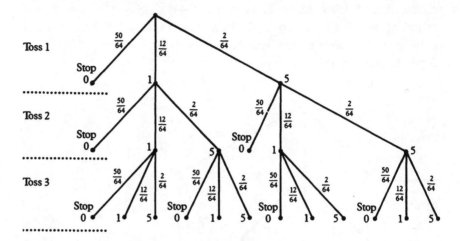

Figure 4.3. Three consecutive tosses—points and probabilities

Table 4.2 summarizes all the possible composite results of one, two, or three consecutive tosses and includes the probability of each result. From the chart we see that chances are almost 5 in 100 that a player gets at least three tosses and about 1 in 100 that a player will also have a fourth toss. Chances of getting more than 5 points are similarly slim, and although chances of getting 15 points on the three tosses are slightly less than 1 in 1000, all of these are possible and can be hoped for by a player.

Table 4.2 Composite results of one, two, and three tosses

Points		Probability	
0	Stop after first toss	78.13% }	= 78.13%
1	Stop after second toss	14.65%	= 17.09%
5	Stop after second toss	2.44%	
2	Stop after third toss	2.75%	
6	Stop after third toss	0.92%	= 3.75%
10	Stop after third toss	0.08%	
3	Continue to fourth toss	0.66%	
7	Continue to fourth toss	0.33%	= 1.04%
11	Continue to fourth toss	0.05%	
15	Continue to fourth toss	0.00%	

That the game holds excitement and interest for players and spectators is clear from its ubiquity among so many different Native American cultures and its persistence through time. In the western portion of North America, it is most often a private game played by women in pairs or small groups. Among the Northeastern Iroquoian peoples, however, it is a public game associated with community ceremonials. In particular, the game concludes certain harvest and New Year's festivals. It is so entrenched and popular a part of the festival tradition that it is believed that the game is played in the future life as well as this one. Another important context for the game of Dish is communal playing to restore to health a person who is ill. The excitement of the game stems not from hopes for material gain or personal recognition but from gaining victory for one's village or clan or nation, or well-being for the sick individual.

Several detailed accounts tell how, upon the request of someone who is ill, an entire village summons another village to join in a game of Dish. The request is made because of a dream. The villagers carefully prepare, some by fasting, some by sexual abstinence, and some by trial games. These activities are undertaken in order to induce dreams that are then compared to decide who shall shake the dish for their side during play. The others will be present, participating through appropriate prayers or chants or invocations. Many goods, collected in each village, are used to bet on the outcome of the play. The goods, large and small, are, in particular, those that have appeared in dreams and so are significant. The games can last for five or six days, sometimes continuing through the nights as well. The atmosphere is excited and noisy, the losses in property can be great, but the game in all its ramifications is intensely serious and spiritually motivated.

The spiritual and communal embedding of this game stands in stark contrast to games of chance as they are most often currently played in Western culture. Some idea of Iroquois beliefs about dreams and illness are necessary for an understanding of the setting for the game of Dish.

The Iroquois recognize as causes of illness natural injuries, witchcraft, and the mind. They believe that lack of mental well-being can lead to illness and can be the cause of physical symptoms. Dreams, in general, are taken very seriously. Recounting them, discussing them, and action related to them are considered of utmost importance. Some dreams, particularly those in which there are visitations from supernatural beings, carry messages for the dreamer or for the entire community. Other dreams are seen as direct or symbolic statements of the desires of the dreamer's soul. For the overall health of the individual, it is urgent that the desires so expressed not be frustrated. When first encountered by Europeans in the 1600s, these Iroquoian ideas were considered strange.

Now, however, with the theories and insights supplied by Freud and his followers, the Iroquois are seen to be remarkably astute. We can now understand that their philosophy incorporates a theory of mind that involves the unconscious, pyschosomatic illness, and dreams as an expression of the unconscious. One anthropologist has noted that the most dramatic dreams are reported for young men approaching puberty rites, warriors getting ready for battle, and people who are ill. These, he says, are stressful situations in which people may well feel in need of guidance and support. Dreams and their public articulation provide a means of expressing these otherwise unspoken needs in a culture that places great emphasis on autonomy, self-reliance, and bravery.

We, in our contemporary Euro-American culture, sometimes read very moving newspaper accounts of a family celebrating Christmas early on behalf of a very sick child. Frequently it is because they fear the child will not be with them when the holiday comes. Mostly, though, it is an extreme statement of love and concern for the child. The parents can no more cause Christmas to come early than the Iroquois can cause the harvest time or New Year to come early. But, by carrying out some of the activities associated with that festival, they can show their willingness to change, if they could, even the yearly round of events. The communal response to a dream request for the game of Dish is to drastically disrupt the normal routine, to lavish attention on the person who is ill, and to willingly risk the loss of even those possessions that are daily necessities. In a culture that recognizes the significant role of the mind in illness, the request for the game of Dish and the fullness and immediacy of the communal response are important means of asking for and giving support when it is needed.

Games of strategy, even more than games of chance, capture our interest as somehow mathematical. If nothing else, the number of mathematicians devoted to these games would suggest a link between them. Although to some a game of chess looks like a royal battlefield, many of its devotees see it more as a challenge in incisive and logical thinking.

In any game of strategy, just as in the creation of a mathematical proof or solution of a mathematical problem, there is a specific end goal. In both, there are a number of alternative legitimate moves from which a particular move must be selected. Each move has to be assessed in terms of what options it opens or closes, and the player has to visualize the effect of a sequence of these moves. Thus, the essential shared feature is thinking through the logical implications of a chain of steps. Further-more, winning in a game or in solving a mathematics problem depends not just on making legitimate moves but on moving steadily toward the goal while so doing. The notion of an elegant play or an elegant proof or problem solution usually refers to a sequence of moves in which none are unnecessary sidetracks or detours; that is, they move sharply and cleanly toward the goal. Another significant feature shared by mathematics problems and games of strategy is implicit in the foregoing statements; all available choices and decisions are confined to the realm of the problem or game at hand. It does not matter whom you know, or what your social skills are, or whether you like the rules of play, or whether you are considered ugly or beautiful.

The major difference between a game and a mathematics problem is that in a game there is an opponent actively trying to frustrate your intentions. That, however, is why a game can be played over and over again. Different opponents, or even the same opponent at different times, can make different responses. Your moves and your opponent's responses interact to create new situations or new problems to be solved.

In attempting to understand what keeps a game interesting, this last point is crucial. Most people familiar with tic-tac-toe recall finding it fascinating at first and then of lessening interest until they stopped playing altogether. What was gradually learned was that for a pair of careful players, the game ends in a draw. The pattern of play became routinized because, with a maximum of only three rounds, the players could easily learn to see the implications of all moves. Some other games, such as nim, also lose our interest when we learn that when both sides play carefully, whoever goes first (or last) is assured of winning.

4 Here we will consider a game of strategy that has been a longtime favorite of the Maori, the indigenous people of the region now called New Zealand. The name of the game is *mu torere*, the etymology of which remains obscure. An 1856 account describes the intensity of the players and the deep interest of the crowd of onlookers. Another account, written more than one hundred years later, includes mu torere as one of the few games that still persists among the Maori despite the upheavals in their culture caused by the over-whelming influx of Europeans and European culture.

The game is simple to describe. The materials needed are few and these are not elaborated in any way. Although the Maori are noted for their wood carving and artistry, the objects used are neither permanent

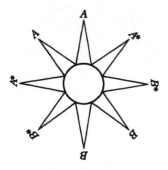

Figure 4.4. Mu torere

nor particularly decorated. Unfortunately, beyond its popularity, very little has been recorded about the cultural context of the game. We know that the players are serious; some people are more expert than others; and the game can be played quietly and privately or can attract a crowd of onlookers.

Mu torere is played on an eight-pointed star design. There are two players and each has four markers. The markers can be pebbles or, after European occupation, bits of broken china or even pieces of potato as long as the players can distinguish them. The star design is easily and readily drawn on sand, wood, or bark with anything that comes to hand such as a piece of charcoal or the point of a nail. The center of the design (see Figure 4.4) is generally referred to as *putahi* and the emanating rays as *kawai*. As with all words, their meaning depends on their context. Other than denoting the parts of this game design, *putahi* means to join or meet as paths or streams running into each other, and *kawai* is variously translated as shoots or branches of a creeping plant, tentaculea of a cuttlefish, or handles of a basket.

Let us call the players *A* and *B* and identify their markers by these letters. To begin the game, the markers are placed as shown in Figure 4.4. Players move alternately, one marker per move. A move can be made to an empty adjacent star point or to the center if it is empty. The game is won when the opponent is blocked so that he cannot move. The blocked player is *piro* which, when associated with a game, translates as "out" or "defeated." (Otherwise, *piro* translates as "putrid" and "foul smelling.") On the first two moves of each player, only their outer markers (identified by asterisks in Figure 4.4) can be moved so that no player is quickly blocked. Some writers report that only the first move of each player is so restricted, but that cannot be the case since the game could then always be won on the second move of the second player. Later we will examine another version of this rule and its implications for the game.

By far, the best way to appreciate this or any game of strategy is to play it. At the beginning, one probably tries, one move at a time, to keep from losing. But as the game goes on, one tries to develop ways to confine the moves of the opponent while simultaneously interfering with his plans for you. So I suggest you play the game, but, in addition, we will analyze it for its underlying structure to better understand what it is that has made the game a Maori favorite for so long.

For the analysis we create a mode of description that includes each possible arrangement of markers on the star and then see how the permitted moves take us from one configuration to another. To become familiar with this mode of description, we first examine a much simplified version of the game. The same rules of moving and winning apply but our simplified version will use a star with just four points and so just two markers per player. Figure 4.5 shows the simplified star and its opening arrangement. Each board descriptor will contain a sequence of six letters. The first letter is the player with the next move (*A* or *B*); the second is the occupant of the center (*A*, *B*, or *O* for no one); and the next four are the markers filling the star points (*A*, *B*, or *O*) read consecutively. If *A* is to go first, the configuration for the opening arrangement shown in Figure 4.6 is *AOAABB*. For the same star layout we could just as well start the reading anywhere on the star (*AOABBA* or *AOBBAA* or *AOBAAB*) or read the points in a counterclockwise direction (*AOBBAA*). Because of the symmetry of the star, these configurations related by rotation or reversal are equivalent. During the analysis we need to use only one from any group that are equivalent. For consistency, we will read the star points clockwise and, when there is an empty star point, we will start with it.

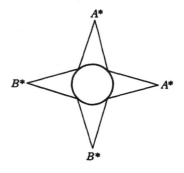

Figure 4.5. Mu torere simplified to a four-pointed star

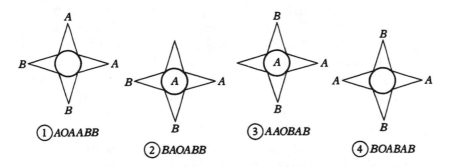

Figure 4.6. Beginning arrangements and their configurations

Now let us trace through the first several moves of this game. Starting with the required opening layout and with *A* having the first move, configuration ① is *AOAABB*. (See Figure 4.6.) The only move that *A* can make is to the empty center and so the new configuration is ② *BAOABB*. *B* goes next and must move to the empty star point; the configuration ③ *AAOBAB* results. *A* has no choice but to move out of the center to ④ *BOABAB*. In all, there are only twelve possible configurations: ① *AOAABB*; ② *BAOABB*; ③ *AAOBAB*; ④ *BOABAB*; ⑤ *ABOABA*; ⑥ *BBOAAB*; ⑦ *ABOAAB*; ⑧ *BBOABA*; ⑨ *AOABAB*; ⑩ *BAOBAB*; ⑪ *AAOABB*; and ⑫ *BOAABB*. Figure 4.7 shows how they are linked together. At the beginning there are no choices: ① leads to ② leads to ③ leads to ④ leads to ⑤ leads to ⑥. Then, depending on what choice *B* makes (at ⑥ it is *B*'s turn to move), the next configuration is ⑦ or back to ① again. The game flow diagram makes clear that there is no way to win this game. For in no configuration is either player blocked from moving, and so the game could go on endlessly. This simplified version of mu torere would not be an interesting game!

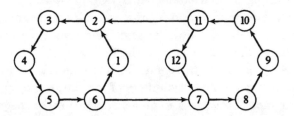

Figure 4.7. The flow of the game on a four-pointed star

Before leaving this version, some further observations can be drawn about the game flow diagram. The diagram is of the same type used in our discussion of the Warlpiri kin system (Chapter 3, section 6); namely, it is a *directed graph*. The board configurations are the vertices and the moves that interconnect them are the directed edges. All possible moves and responses are summarized in the graph. If *B* were to make the opening move, the game would start at ⑫ and then proceed to ⑦ to ⑧ and so on. Further, since who goes first should not affect the logical flow of the game, each board configuration has a complementary configuration in which the *A*'s and *B*'s are interchanged, and these complements are similarly linked. On the game flow diagram the complementary configurations are those that are symmetric with respect to the center point of the diagram (① and ⑫, ② and ⑦, ③ and ⑧, and so on).

Next, we will complicate this simplified game by using three markers per player and a six-pointed star. (We still, of course, are not at the game of mu torere.) Now each board configuration is eight letters: who goes next (*A* or *B*); who is in the center (*A*, *B*, or *O*); and the six consecutive star point markers. There are, in all, thirty possible board configurations. They are listed in Figure 4.8 and the flow of the game is shown in Figure 4.9. If *A* goes first, the game starts at ① and if *B* goes first it starts at ㉕. The dotted lines are the moves not permitted on the first two rounds. The dotted paths from ① to ③ and from ㉕ to ㉖ show that, were it not for this rule, the game could end on the first round. (The restriction on the second round is not pertinent in this version.) This game could be enjoyable. If *A* begins and *B* is a very poor player, *A* could win on the fourth round.

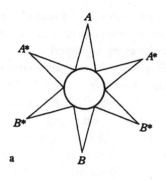

① *AOAAABBB*	⑪ *AAOBABBA*	㉑ *AOBABABA*
② *BAOBBBAA*	⑫ *BAOBBABA*	㉒ *BAOBABAB*
③ *BAOABBBA*	⑬ *AAOBABAB*	㉓ *BAOBAABB*
④ *AAOBBAAB*	⑭ *BOABABAB*	㉔ *AAOBBBAA*
⑤ *BOBBAABA*	⑮ *ABOABABA*	㉕ *BOBBBAAA*
⑥ *ABOABBAA*	⑯ *BBOAABAB*	㉖ *ABOBAAAB*
⑦ *BBOAAABB*	⑰ *ABOBAABA*	㉗ *ABOAAABB*
⑧ *BBOAABBA*	⑱ *BBOABAAB*	㉘ *AAOBBABA*
⑨ *AOAABBAB*	⑲ *ABOAABAB*	㉙ *AAOABBBA*
⑩ *BAOABBAB*	⑳ *BBOABABA*	㉚ *BBOBAAAB*

Figure 4.8. Mu torere simplified to a six-pointed star

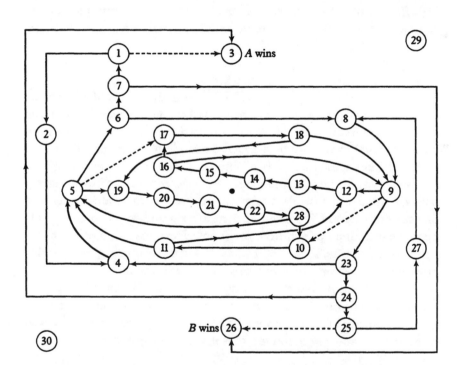

Figure 4.9. The flow of the game on a six-pointed star

If this were *A*'s plan, with a little care *B* could win even sooner via

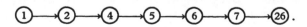

But *A* could frustrate this strategy by going from ⑥ to ⑧ rather than ⑥ to ⑦, thus opening new challenges and new possibilities for both

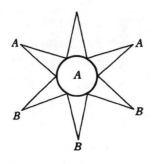

Figure 4.10. An arrangement in which *B* is blocked

of them. In any case, there are ways to win and pitfalls to avoid along the way. Regardless of who starts, neither player wins automatically, so winning or losing depends on how carefully the game is played.

Notice on the game flow diagram (Figure 4.9) the anomalous configurations ㉙ and ㉚ (upper right and lower left). They exist in theory but can never be attained in actual play. In graph theoretic terms, they are vertices of the graph of the game but, because no edges connect to them, they are *isolated vertices*. We can see what characterizes them by comparing them to the winning configurations. There is no way for *B* to move out of the star arrangement of Figure 4.10. Similarly, there is no way for *B* to have moved to create it. Thus, when this arrangement results from a move by *A*, it is *A*'s winning configuration ③. But it cannot result from a move by *B*; that is the implication of configuration ㉙. Each winning configuration for *A*, then, has a counterpart that is unattainable, as does each winning configuration for *B*.

5 From these simplified versions of mu torere with just two or three markers per player on four- or six-pointed stars, we now move to an analysis of the considerably more intricate actual game. We have in hand a mode of describing the board configurations, some idea of their different possible types, and a method of displaying the flow of the game. However, instead of haphazardly trying to list all the possible configurations, we first calculate the number that should be found. This will be done in a general way so that it builds upon the intuitions gained from the simpler versions and uses them as a check on the results. In addition, considering the problem generally also provides a framework for generating the configurations.

To cast the problem generally, let n be the number of pieces for each player. The number of star points is then $2n$ and a board configuration is a sequence of $2n + 2$ letters. (We have already examined the special cases of $n = 2$ and $n = 3$; mu torere is $n = 4$.) Our questions are:

a. How many winning configurations are there for each player?
b. How many configurations are theoretically possible but never attained in play?
c. How many total configurations are there with the center filled?
d. How many total configurations are there with the center empty?

(If you are wary of general formulations because of the symbolism they necessarily involve, you can rejoin us in section 6. By then we will have these questions answered and will have evaluated the resultant formulas for $n = 4$.)

As in most problems that involve arrangements of things, the concept and notation of *combinations* are useful. Essentially, combinations address the *number* of ways items can be combined together to form certain arrangements. If, for example, there were seven items to be placed in order, the number of different ordering would be $7 \cdot 6 \cdot 5 \cdot 4 \cdot 3 \cdot 2 \cdot 1$. That is, the first position could be filled with any one of the seven items. Then the next could be any one of the other six. Together that would give 42 different possible arrangements. Continuing on, each of these 42 arrangements could then have any one of the remaining five items appended giving now $7 \cdot 6 \cdot 5$ possibilities, and so on. This product of $7 \cdot 6 \cdot 5 \ldots 1$ is shorthanded with the symbol *7!* (read *seven factorial*). If, however, some of the items are indistinguishable from one another, some of the orderings are identical to each other but have been counted as different. To compensate for the redundancy, the answer is divided by the multiplicity introduced by the indistinguishable items. If, say, four items were indistinguishable from one another, they would have introduced a multiplicity of $4 \cdot 3 \cdot 2 \cdot 1$ (that is, 4!) because that is the number of ways four *different* items can be arranged. So, in the example of seven items, if four were of one indistinguishable category and the remaining three were also indistinguishable but of another category, the number of different orderings would be

$$\frac{7!}{4!3!} = \frac{7 \cdot 6 \cdot 5 \cdot 4 \cdot 3 \cdot 2 \cdot 1}{(4 \cdot 3 \cdot 2 \cdot 1)(3 \cdot 2 \cdot 1)} = 35.$$

This computation is denoted by the symbol $\binom{7}{4}$ and is read as *the number of combinations of seven things four at a time*, or as just *seven chose four*. (These words are associated with the symbol because of a somewhat different application that results in the same computation.) Since 4 and 3 add to 7, $\binom{7}{3}$ denotes the same computation but focuses on the group of three. In general,

$$\binom{n}{r} = \binom{n}{n-r} = \frac{n!}{r!(n-r)!}$$

where n is the total number of items, r is the number of indistinguishable items in one category, and the remaining $n - r$ are also indistinguishable but in another category. $n!$ means $n \cdot (n-1) \cdot (n-2) \ldots 1$, with the special proviso that $0! = 1$. In our case the items being ordered are all either A's or B's. Hence, when we count different orderings, we will always be dealing with some total number of items made up of some A's and the rest B's.

a. Winning Configurations In order for A to win, B must be blocked from moving. This can occur only when it is B's turn to move, the center is filled by A, and the empty position on the star has A on both sides of it. Therefore, a winning configuration for A must be $BAOA \ldots A$ with $2n - 3$ positions between the A's. The $2n - 3$ positions contain $n - 3$ A's and n B's. There are $\binom{2n - 3}{n}$ different possible orderings when n is 3 or more but no possibilities for an n of 2 or 1.

As we noted earlier, orderings that are due to reading star points clockwise and counterclockwise are, for our purposes, equivalent and so only one of them should be counted. Were each counterclockwise reading different from its related clockwise reading, the total count we already have could simply be divided by two. For some star layouts, however, because of the symmetry of the placement of A's and B's, both readings give an identical result; these are in the count just once. Before dividing by two then, we must find and compensate for the number of these self-reversals.

Since each possible ordering has n B's and $n - 3$ A's, when n is odd there are an odd number of B's and an even number of A's. A self-reversal must have symmetry around its midpoint and so must have a B in the center and $n - 2$ items (that is, half of the remaining items) in reverse order on either side. The $n - 2$ items consist of $(n - 2)/2$ A's and $(n - 1)/2$ B's. The number of such arrangements is $\binom{n - 2}{(n - 1)/2}$. On the other hand, when n is even, there are an odd number $(n - 3)$ of A's so an A must be central. The $n - 2$ items on each side are then $(n - 4)/2$ A's and $n/2$ B's. The number of such arrangements is $\binom{n - 2}{n/2}$. Using $W(A)$ to denote the number of winning configurations for A, a summary of these results is:

$$W(A) = \begin{cases} \frac{1}{2}\left[\binom{2n - 3}{n} + \binom{n - 2}{(n - 1)/2}\right] & \text{for odd } n \text{ of 3 or more} \\ \frac{1}{2}\left[\binom{2n - 3}{n} + \binom{n - 2}{n/2}\right] & \text{for even } n \text{ of 4 or more} \\ 0 & \text{for } n = 1 \text{ or 2.} \end{cases}$$

The winning configurations for B are the complements of the winning configurations for A so these too number $W(A)$.

b. Unattainable Configurations As was already illustrated, every winning configuration has a counterpart that is unattainable. They differ only in the first letter indicating the player with the next move. Winning configurations for A have the form $BAOA \ldots A$ and so any of the forms $AAOA \ldots A$ is unattainable. Conversely, any unattainable configurations with A as the first letter must be of this form. Thus, their number must equal $W(A)$. Similarly, the unattainable configurations with B as the first letter are counterparts of winning configurations for B. Thus, these too number $W(A)$.

c. Total Number of Configurations with Center Filled Let us first concentrate on configurations in which the center is filled by A and A has the next move. These are of the form $AAO \ldots$ where the remaining $2n - 1$ positions contain n B's and $n - 1$ A's. There are $\binom{2n - 1}{n}$ different orderings possible. As before, this total count must be corrected for the equivalent clockwise and counterclockwise orderings. Using the same reasoning as before, the number of self-reversals is $\binom{n - 1}{n/2}$ if n is even or $\binom{n - 1}{(n - 1)/2}$ if n is odd. Denoting the resulting number of configurations of this type by $N(AA)$:

$$N(AA) = \begin{cases} \frac{1}{2}\left[\binom{2n - 1}{n} + \binom{n - 1}{n/2}\right] & \text{for } n \text{ even} \\ \frac{1}{2}\left[\binom{2n - 1}{n} + \binom{n - 1}{(n - 1)/2}\right] & \text{for } n \text{ odd.} \end{cases}$$

The same number of configurations would result for B in the center with A having the next move and for each A or B in the center with B having the next move. Thus, the total number of configurations with center filled is $4N(AA)$.

d. Total Number of Configurations with Center Empty When A has the next move, the configurations with empty centers are $AO \ldots$ where the remaining $2n$ positions contain n A's and n B's. The problem is more difficult than it appears at first glance. In the foregoing situations, there was an empty star point that could be used for orientation. Now, without that, the circular nature of the layout must be considered. There are now equivalent readings due to rotations as well as reversals. The difficulty is that some rotations replicate each other and some do not, just as some reversals replicate each other and some do not. In addition, some reversals replicate, not themselves, but rotations of themselves. In total, $2n$ markers can be arranged in $\binom{2n}{n}$ different ways. This count must be corrected for the repetitions due to rotation, the repetitions due to reversals, and also for the repetitions due to reversals of rotations. Rather than stating the details, I will just state the result, explain what it says, and then, in the notes, refer to detailed treatments of the problem.

The number of configurations when A has the next move and the center is empty (and which are *inequivalent* under rotation and reversal) is:

$$N(A) = \begin{cases} \dfrac{1}{2}\left[\dfrac{1}{2n}\sum_{d|n}\varphi(d)\binom{2n/d}{n/d} + \binom{n}{n/2}\right] & \text{for } n \text{ even} \\[4mm] \dfrac{1}{2}\left[\dfrac{1}{2n}\sum_{d|n}\varphi(d)\binom{2n/d}{n/d} + \binom{n-1}{(n-1)/2}\right] & \text{for } n \text{ odd.} \end{cases}$$

$\varphi(d)$ is the *totient function*. It is the number of positive integers less than d and having no factors in common with d. For example, the integers less than 6 that have no factors in common with 6 are only 1 and 5. Therefore, $\varphi(6) = 2$. As another example, the integers less than 7 having no factors in common with 7 are 1, 2, 3, 4, 5, 6 so $\varphi(7) = 6$. The symbol $\sum_{d|n}$ means that each positive integer that divides evenly into n is to be used as d and then the results summed. If n is, say, 3, its divisors are 1 and 3 so the values $\varphi(1)\binom{6}{3}$ and $\varphi(3)\binom{2}{1}$ are summed where $\varphi(1) = 1$ and $\varphi(3) = 2$. For n equals 4, the divisors are 1, 2, and 4 so $\varphi(1)\binom{8}{4}$ and $\varphi(2)\binom{4}{2}$ and $\varphi(4)\binom{2}{1}$ are summed where $\varphi(1) = 1$, $\varphi(2) = 1$, and $\varphi(4) = 2$.

Table 4.3. Summary of configurations

	$n=2$	$n=3$	$n=4$	$n=5$	Formula
Total configurations	12	30	92	296	$4N(AA) + 2N(A)$
Center filled	8	24	76	264	$4N(AA)$
Center empty	4	6	16	32	$2N(A)$
Wins for A	0	1	3	12	$W(A)$
Wins for B	0	1	3	12	$W(A)$
Unattainable	0	2	6	24	$2W(A)$

When B has the next move with centers empty, the configurations are $BO\ldots$, again with the remaining $2n$ positions containing n A's and n B's. The problem is the same and so the number of them is also $N(A)$.

We are now ready to accumulate and summarize our results and evaluate them for a few specific n. See Table 4.3.

6 In all, there are 92 configurations for the game of mu torere. Six of them, three for each player, are winning configurations and another six can never be attained. A complete list is shown in Table 4.4 and Figure 4.11 is the game flow diagram. The

Table 4.4. Configurations for mu torere

① *AOAAAABBBB*	㉔ *AAOABBBAAB*	㊼ *BOBBBBAAAA*	(70) *BBOBAAABBA*
② *AOAAABABBB*	㉕ *AAOABABBBA*	㊽ *BOBBBABAAA*	(71) *BBOBABAAAB*
③ *AOAAABBABB*	㉖ *AAOABBABBA*	㊾ *BOBBBAABAA*	(72) *BBOBAABAAB*
④ *AOAABAABBB*	㉗ *AAOABBBBAA*	㊿ *BOBBABBAAA*	(73) *BBOBAAAABB*
⑤ *AOAABBAABB*	㉘ *ABOAAABBBA*	(51) *BOBBAABBAA*	(74) *BAOBBBAAAB*
⑥ *AOAABABABB*	㉙ *ABOBABAAAB*	(52) *BOBBABABAA*	(75) *BAOABABBBA*
⑦ *AOAABABBAB*	㉚ *ABOABBAABA*	(53) *BOBBABAABA*	(76) *BAOBAABBAB*
⑧ *AOABABABAB*	㉛ *ABOAABBABA*	(54) *BOBABABABA*	(77) *BAOBBAABAB*
⑨ *AAOBBBAAAB*	㉜ *ABOBBABAAA*	(55) *BBOAAABBBA*	(78) *BAOAABABBB*
⑩ *AAOAABBBAB*	㉝ *ABOABAAABB*	(56) *BBOBBAAABA*	(79) *BAOBABBBAA*
⑪ *AAOBABAABB*	㉞ *ABOBABABAA*	(57) *BBOABABBAA*	(80) *BAOABABABB*
⑫ *AAOBAABBAB*	㉟ *ABOABABAAB*	(58) *BBOABBAABA*	(81) *BAOBABABBA*
⑬ *AAOBABABBA*	㊱ *ABOBAAABBA*	(59) *BBOABABAAB*	(82) *BAOABBBAAB*
⑭ *AAOAABABBB*	㊲ *ABOAAABBAB*	(60) *BBOBBABAAA*	(83) *BAOBBBAABA*
⑮ *AAOABABABB*	㊳ *ABOAABABBA*	(61) *BBOBABABAA*	(84) *BAOBBABAAB*
⑯ *AAOBBBAABA*	㊴ *ABOBAABAAB*	(62) *BBOAAABBAB*	(85) *BAOABBABBA*
⑰ *AAOBBBBAAA*	㊵ *ABOAABAABB*	(63) *BBOAAAABBB*	(86) *BAOBBABBAA*
⑱ *AAOBBAAABB*	㊶ *ABOAABBBAA*	(64) *BBOAABBBAA*	(87) *BAOBBAAABB*
⑲ *AAOBABABAB*	㊷ *ABOBAAAABB*	(65) *BBOABABABA*	(88) *BAOABBBBAA*
⑳ *AAOABABBAB*	㊸ *ABOAAAABBB*	(66) *BBOBABAABA*	(89) *BAOBBBBAAA*
㉑ *AAOBBABBAA*	㊹ *ABOABABABA*	(67) *BBOAABAABB*	(90) *BAOBABABAB*
㉒ *AAOABBAABB*	㊺ *ABOAABBAAB*	(68) *BBOBAABBAA*	(91) *BAOBBAABBA*
㉓ *AAOBAABABB*	㊻ *ABOBABAABA*	(69) *BBOABBABAA*	(92) *BAOABABBAB*

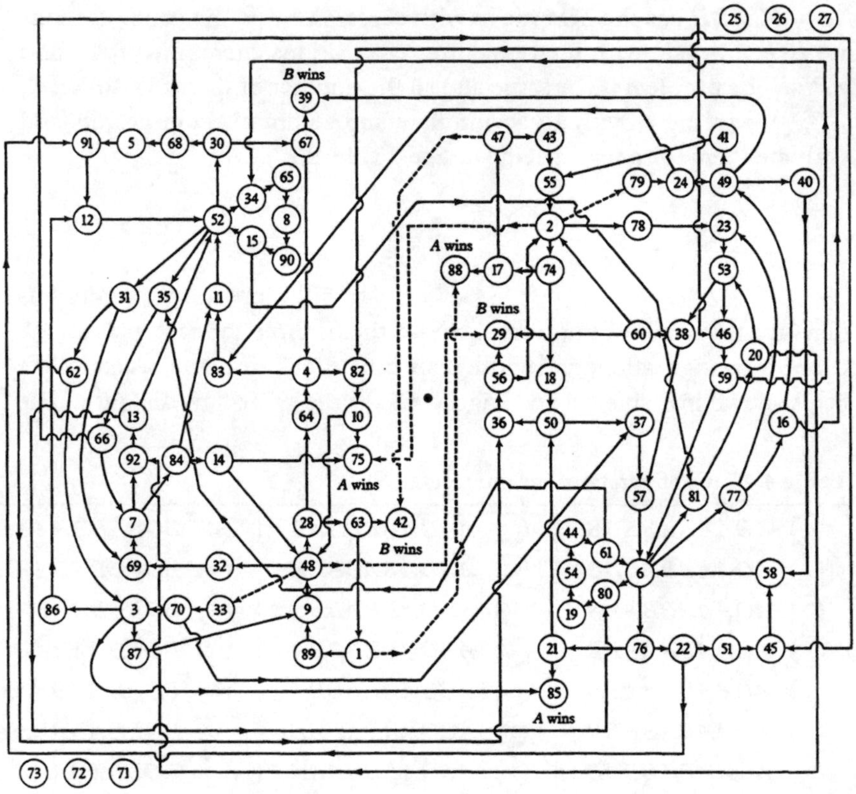

Figure 4.11. The flow of the game of mu torere

game begins at ① or ㊼; the dotted paths are those not permitted on the first two rounds. If *A* goes first, the opening sequence must be ①→(89)→⑨→㊽. Then the choices begin. Since it is *B*'s second move, only ㊽→㉘ or ㊽→㉜ are possible. The former would not be wise as *A* could then easily win with ㉘→(63)→①→(88). Even using the latter, *B* must soon again exercise caution as ㉜→(69)→⑦→(84)→⑭→(75) is a win for *A*. *A* must be cautious as well. For example, choosing ㉛→(62) rather than ㉛→(69) would lead to a win for *B*, or choosing ④→(83) rather than ④→(82) misses winning the game in a few more moves. There are, for either player, short routes to winning against an inexperienced or careless opponent but, with two careful players, the game can go on indefinitely. The many points of choice and multiple winning configurations make for great variety and require skill to avoid approaching pitfalls while planning how to entrap one's opponent.

7 Just as we looked at simpler versions of mu torere with fewer markers per player, we briefly look at versions with more. The result is surprising: the games would *not* be enjoyable. There are many more configurations, but if the first player is careful, he can win in the fourth round. And if he is careless, his opponent can win in the third round.

Just tracing through a few moves for the game with five markers per player on a ten-pointed star can show why this happens. Only some of the 269 configurations and the relevant part of the game flow diagram are shown in Figure 4.12. After the first two rounds, three *A*'s are still adjacent. The middle one can be moved to the center, as soon as the center is free, without leaving a star point empty that is next to an opponent. Then, of course, the opponent is blocked and the game is won. For more than five pieces per player, the lineup of *A*'s would be even greater than three and so the same strategy will work. Mu torere with its four markers per player on an eight-pointed star is, therefore, the most enjoyable version of this game.

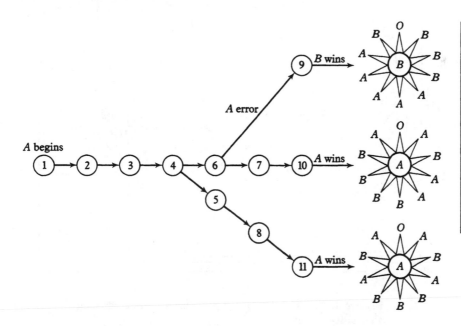

1. *AOAAAAABBBBB*
2. *BAOBBBBBAAAA*
3. *AAOBBBBAAAB*
4. *BOABBBBAAAAB*
5. *ABOAAAABABBB*
6. *ABOABBBBAAAA*
7. *BBOAABBBBAAA*
8. *BBOAAABABBBA*
9. *BBOBBBBAAAAA*
10. *AOABBBBAAABA*
11. *AOABABBBABAA*

Figure 4.12. The flow of the game on a ten-pointed star

These unenjoyable enlarged versions further highlight the importance of the mu torere rule that restricts the two opening moves. The restriction forces dispersal of the markers. As was mentioned at the very start of our discussion, some accounts cite a variation on this rule. The variation is that *throughout the game* no marker that is between two like markers can be moved. Thus, a player cannot benefit should the dispersed markers become adjacent again. Figure 4.13 shows the game flow diagram modified to reflect this. Comparison with the flow diagram of the game under the more common rule (Figure 4.11) shows the same attainable configurations and the same winning configurations. There are, however, fewer routes to the winning configurations. As a result, I suspect, the players are faced with a greater challenge to their strategic skills.

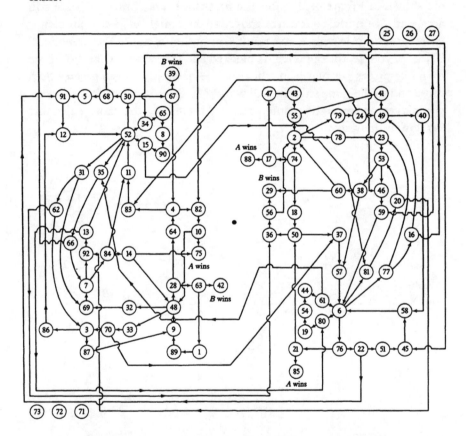

Figure 4.13. The flow of the game of mu torere when an alternative rule is used

Indigenous games of strategy are rare among Oceanic people. Mu torere is one of the few that is known to exist. For us it has raised some combinatorial questions that are interesting in and of themselves. It has also stimulated us to seek a way to exhibit and explore the interplay of the specific rules that define the game. Our purpose has been to better understand what has made mu torere such a longtime favorite. In any case, the game is testimony to the fact that logical play is not confined to any particular peoples.

8 Logical puzzles are closely akin to games of strategy but usually require much shorter paths to success. Some puzzles, posed in story form, have become such favorites that many of us cannot even recall when we first heard them. As contrasted to a game, the story puzzle has no objects to be manipulated and no opponent with whom to contend. Nevertheless, it presents a logical challenge because the story sets a goal and specifies constraints on how the goal can be achieved. Once solved, further enjoyment comes from posing the problem to others or in creating variations that add a slight twist to the logic. The very nature of these puzzles, namely that they are conveyed through stories, embeds them in the culture in which they are found. A story, to be understood at all, must in some way reflect or express situations or ideas that have meaning to the listeners.

A longtime favorite in Western culture is a puzzle involving six people who must cross a river in a boat that can hold only two at a time. The story has many different versions; the essence of each, however, is always who may or may not be left alone with whom. Jealous husbands fear adultery will be committed if their wives are with other men; servants fear being beaten by masters other than their own; missionaries fear cannibals; merchants fear robbers; masters fear slave rebellions; and in-laws fear a quarrel. These fears provide the logical constraints but they also express our changing views of who is afraid of whom and why. The mathematical nature of the problem is underscored by the fact that, although couched in real-world terms, one is not allowed to resort to an open variety of realistic solutions. You cannot seek another boat, use a bridge further up the river, or patch up the dispute between in-laws. The story not only poses the problem but also clearly delimits the means of its solution.

Another river-crossing puzzle has as long a history and an even greater spread. Found with variations in African cultures as well as in Western culture, this logical puzzle is yet another example of the wide-spread enjoyment of logical challenges. In its most familiar form, a man must ferry across a river a wolf, a goat, and a head of cabbage. The difficulty is that the available boat can only carry him and one of these but neither the wolf and goat nor the goat and cabbage can be left alone together.

The Western origin of the puzzle is most often attributed to a set of 53 problems designed to challenge youthful minds, "Propositiones ad acuendos iuvenes." The problem set was circulated around the year 1000, but the author is said to be Alcuin of York, a theologian, writer, and teacher who lived from 735 to 804, as he referred to them in a letter to his most famous student, Charlemagne. The solution given by this work is to carry over the goat, then transport the wolf and return with the goat, then carry over the cabbage, then carry over the goat. A second solution, which simply interchanges the wolf and cabbage, is often attributed to the French mathematician Chuquet in 1484 but is found even earlier in the twelfth century in Germany in the succinct form of Latin hexameter: *It capra, fertur olus, redit hec, lupus it, capra transit.*

The exact authorship is less important than the fact that the problem has circulated for over a thousand years, both orally and in writing. Not only is the puzzle repeated over and over again in mathematical recreation books, it is also found as a folk puzzle by collectors of folklore. I have read the puzzle in collections of Gaelic, Welsh, Danish, Russian, Italian, Rumanian, Saxon, and African-American folklore. In the French Brittany region, the puzzle is just one component of a lengthy tale about Jean L'Hébété (the dazed or simpleminded) who is eventually given more intelligence by a good fairy in exchange for his wife's solution to her river-crossing predicament. Here, it is the combination of love and logic that solve the couple's problem. Sometimes the characters are changed—a fox replaces the wolf, a sheep or buck replaces the goat, or hay replaces the cabbage. And, as the puzzle passes from Europe to America, the goat and cabbage give way to a fowl and some corn; corn, of course, is indigenous to the environment and was introduced to Europeans by Native Americans.

Despite the differences, with one exception, all of the Western versions share the same logical structure: A, B, C must be transported across a river in a boat that can only hold the human rower and one of A, B, C; neither A nor C can be left alone with B on either shore. (We return to the exception later as it is significant in its relation to another part of the world.) The solution is shown in Figure 4.14.

Quite the same puzzle, with a wolf, a goat, and cabbage, is found as a folk story on the Cape Verde Islands just off the Western coast of Africa. Among the Tigre of Ethiopia, a leopard is the predator and a leaf is the food. And, for the Bamiléke of Cameroon, the water is only a stream

	Side 1	In transit	Side 2
	Man, A, B, C	—	—
Round trip 1	A, C	Man, $B \longrightarrow$ $\longleftarrow$ Man	— B
Round trip 2	A	Man, $C \longrightarrow$ $\longleftarrow$ Man, B	B C
Round trip 3	B	Man, $A \longrightarrow$ $\longleftarrow$ Man	C A, C
Last trip	—	Man, $B \longrightarrow$	A, C
	—	—	Man, A, B, C

Figure 4.14. Transporting A, B, and C. The two statements of the solution reflect the interchange of the wolf and food as A or C.

and the means of crossing is walking on a fallen tree trunk but, here too, a tiger, a sheep, and a big spray of reeds have to be taken to the other side. Versions of the problem, found in folk stories in three other regions of Africa, are, at first glance, quite similar. They too require a human to transport across some water a predator, its prey, and some food. Closer examination, however, shows that they have distinctly different logical structures. What is more, all three are variations on a single logical form.

The shared structure of these three African story puzzles is: A, B, and C must be transported across a river by a human who can only transport *two* of A, B, C at one time. Neither A nor C can be left alone with B on either shore. This puzzle has more solutions than the Western form; just two of the solutions are shown in Figure 4.15. It is the details of each story that rule out some solutions and make others acceptable. Another unifying characteristic that further separates them from the Western ones is the mode in which they are set forth. The stories are more discursive; that is, they not only set the constraints but elaborate on their ramifications and on the solutions.

Let us look first at the puzzle as it is found among the Kabjlie who live in the Djurdjura mountains on the coast of Algeria. A man must cross a river with a jackal (A), goat (B), and bundle of hay (C). His solution, however, is neither of those we just showed in Figure 4.15. It is shown in Figure 4.16. Another traveler, seeing this, calls to his attention that this solution is less efficient than what we have called solution 1 (Figure 4.15) because the goat is carried on all trips. "Or," he adds, "did you think that jackals eat hay?" Thus, in the exchange, the story goes beyond just stating a puzzle and a solution; a second solution is introduced and contrasted to the first. The traveler is pointing out that, as he sees it, a *good* solution should be concerned not only with the *number* of trips but with the lightest load on each trip. And, furthermore, he is highlighting the fact that if A cannot be alone with B and B cannot be alone with C, it does not necessarily follow that A cannot be alone with C. In

Solution 1

	Side 1	In transit	Side 2
	Man, A, B, C	—	—
Round trip 1	B	Man, $A, C \longrightarrow$ $\longleftarrow$ Man	— A, C
Last trip	—	Man, $B \longrightarrow$	A, C
	—	—	Man, A, B, C

Solution 2

	Side 1	In transit	Side 2
	Man, A, B, C	—	—
Round trip 1	A, C	Man, $B \longrightarrow$ $\longleftarrow$ Man	— B
Last trip	—	Man, $A, C \longrightarrow$	B
	—	—	Man, A, B, C

Figure 4.15. Solutions 1 and 2 to transporting A, B, and C in a boat that can hold two of them

Solution 3

	Side 1	In transit	Side 2
	Man, A, B, C	—	—
Round trip 1	C	Man, $A, B \longrightarrow$ $\longleftarrow$ Man, B	— A
Last trip		Man, $B, C \longrightarrow$	A
	—	—	Man, A, B, C

Figure 4.16. Solution 3 to transporting A, B, and C

this version of the story there is no boat; the river is sufficiently shallow so that the man can walk across carrying one of the objects under each arm. While this does not seem to affect the logic of the puzzle, comparison with the Kpelle version shows that it does indeed.

The Kpelle puzzle is elaborated in a lengthy story. Collected in the northern part of the Kpelle regions on the west coast of Africa (in Liberia), the story is set in the kingdom of King Tokolo near the Pauls River. Among the Kpelle, when a couple marry, the man pays a bride price to the woman's father. This is sometimes waived if the bride's parents are particularly fond of their prospective son-in-law. The story begins when a young man asks for the hand of the king's daughter. Since she is agreeable to the match, the king will forgo the bride payment if the suitor shows himself to be one of the "smart people." To do so, of course, he must meet a challenge set by the king. The king has caged a cheetah and fed it fowls so that it now grabs and eats any fowl near it. The suitor must transport the cheetah, a fowl, and some rice across the river in a boat that holds one person and two of these. But, the king points out, the man cannot control them while rowing the boat, and so in addition to the cheetah and fowl or fowl and rice not being alone together on either shore, they also cannot be together on the boat. The young man tries various solutions and has to appeal to his father for replacement of the fowl and rice when the solutions fail. While the father helps in this way, he cannot help to resolve the puzzle. As we read the story, it is necessary to bear in mind that a theme in Kpelle culture is individual achievement; most important is that achievement should be through personal effort and shrewdness. Thus, the Kpelle father warns his son that he must solve the problem by himself because otherwise he will be shamed, then laughed at, and then become bitter toward himself. Eventually, after the young man succeeds by using solution 1, both families join together for a joyful wedding ceremony.

The third puzzle that shares this logical structure but has its own distinctive twist is a Swahili version. Again it is part of a lengthy story of trials, a mode that is common in African folk tales. Swahili is a Bantu lingua franca spoken by as many as 40 million people throughout a large part of East and Central Africa. It is the indigenous language only in the area including the islands of Zanzibar and Pemba and the facing coast. The story, told at the turn of the century by Ukami and Kuthu visitors to the coast whose home language was not Swahili, is set in a sultanate such as was found in Zanzibar until the late 1800s. The sultan has covered the floor and ceiling of his court with mirrors, and court visitors must pay a tribute for the spectacle of seeing people above and below the room. A visitor from another region refuses to pay, and the sultan confronts him with a challenge. If he can carry a leopard, a goat, and some tree leaves to the sultan's son, tribute will be given him and he can remain in the sultanate. The price of failure is death. The son, of course, lives across a river and the available boat can hold only the traveler and two of the items. This time the added caveat, as stated by the traveler, is that *no* two things can be left alone together on a shore. Thus, of necessity, his solution is like that of the Kabjlie (solution 3) with the leaves as *A* and the leopard as *C*; that is, he first carries over the leaves and goat, returns with the goat, and then carries over the goat and leopard. Again, logic triumphs and the traveler is made welcome in the land.

It is right here that the single Western exception to the Western puzzle form becomes significant. Found among African-Americans on the Sea Islands of South Carolina, it is another of this same African logical form. The characters are the typical American ones—a fox, a duck, and some corn—but two of them can be fit into the boat. The solution is somewhat like solution 3; this time, however, it is the corn that is carried back and forth on each trip.

There is yet one more African version of the river crossing problem distinct in logical form from both the Western ones and the African ones just discussed. This story is found only among the Ila of Zambia. Its striking difference is that there are four items to be transported: a leopard, a goat, a rat, and a basket of kafir corn (sorghum). The boat can hold just the man and one of these. The change from three to four items makes the problem insoluble by logic alone and surrounds it with an ethical dilemma. Here, the interplay of culture and logical constraints becomes most apparent; a choice that might be simple for us and would reduce the problem to its familiar solvable form, is unacceptable to the Ila. The Ila, first of all, have a relationship with plants and animals that is quite different from our own. They believe in metempsychosis and temporary metamorphosis; that is, at death or temporarily during life, a person may pass into another living creature or a plant. A further bond is that clans are totemistic, with plants, animals, or natural objects as totems. (Each clan is named for and believes itself to be in a special relationship to its totem.) And, responsibility when on a journey is of particular importance; letting harm befall a member of your group is a serious offense. It is specified in Ila law that if you ask someone to go somewhere or take someone in a canoe and he meets with any accident, you are guilty of *buditazhi* and can be seized and held for ransom. In short, while for us deserting a rat might be easy, for the Ila traveler it is another matter. After considering sacrificing the rat or sacrificing the leopard, the man's decision is that since both animals are to him as children, he will forgo the river crossing and remain where he is!

The broad geographic spread of this river crossing puzzle and the fact that it appears in three distinctly different logical forms is, in itself, an intriguing problem. (The African sites are shown in Map 3.) There is no demonstrated historical connection among the three forms. We do not even know the exact connections among the variations within each form. While speculation is possible, without information on the first appearances of each version in each place, no hypothesis can be substantiated. What we can be sure of, however, is that numerous different peoples have made this puzzle their own. The need to get unmanageable things across some water may seem fanciful if viewed from a twentieth-

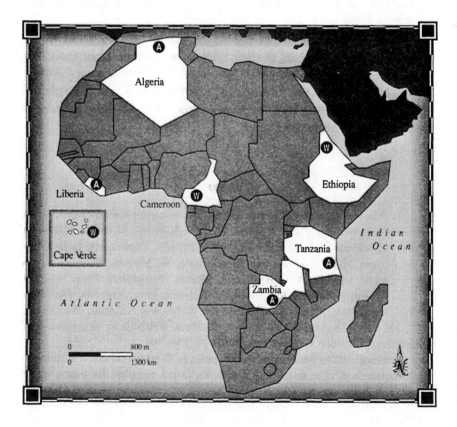

Map 3. Sites of the river crossing puzzle in Africa. *W* is what we have referred to as the Western form; *A* is the African form; and *A′* is the version with four items.

century urban setting. But the problem is not uncommon today in other settings, and surely was not uncommon during the last thousand years. Abstracted from a common experience, stated as a logical challenge, embedded in a story made meaningful through local variations, this story puzzle is testimony to the widespread enjoyment of logical play.

9 We began this chapter discussing games in general and then focused quickly and specifically on games of a mathematical nature. Games of chance and games of strategy are different in kind; the former implicitly involve probabilistic ideas and the latter deal with logical choices and their implications. Both, however, are circumscribed by formal rules and have clearly defined goals. While games may be abstracted from experience or useful for increasing certain skills or be parts of other cultural expressions, what goes on in a game is confined to the realm of the game.

Games of another culture provide us with an unusual window into the ideas of others; we can, in some limited sense, experience them. We cannot, for example, try out another culture's kin system or world view. Although the context and ambiance of a game are different if we play it, we can at least participate in its formal aspects and so share in some of the same mental processes.

Notes

1. The classification of games and the idea that games are models of social interaction are based on J. M. Roberts, M. J. Arth, and R. R. Bush, "Games in culture," *American Anthropologist*, 61 (1959) 597–605. There is a wide literature on the African Mancala game. See, for example, M. Nsimbi, *Omweso, a game people play in Uganda*, UCLA African Studies Occassional Paper 6, 1968; L. Russ, *Mancala Games*, Reference Publications, Inc., Algonac, Michigan, 1984; C. Zaslavsky, *Africa Counts*, Prindle, Weber and Schmidt, Inc., Mass., 1973, pp. 116–136; and P. Townsend, "African Mankala in anthropological perspective," *Current Anthropology*, 20 (1979) 794–796.

Although divination is surely *not* a game, in Western culture it is linked to the forerunners of dice as gaming pieces. In Greek temples, for example, the throwing of five astragali (foot bones of hoofed animals) was a means of divination. The sides of each were identified with one, two, three, and four and different resultant combinations were associated with different Gods and different prophesies. The throw of Zeus was:

One one, two threes, two fours,
The deed which thou meditatest, go, do it boldly.
Put thy hand to it. The gods have given thee favorable omens.
Shrink not from it in thy mind. For no evil shall befall thee.

For a discussion of early dice and divination, see pp. 1–27 in *Games, Gods and Gambling*, F. N. David, Griffin & Co., London, 1962 or her quite similar "Dicing and gaming (a note on the history of probability)," *Biometrika*, 42 (1955) 1–15. According to her, the Zeus throw statement I have quoted is a translation by Sir James Frazer of a Greek inscription. The *I Ching*, originated by the Chinese, is another case involving throwing objects (six sticks) in order to locate a portion of text that provides counsel. The combinatorial basis of the *I Ching* is discussed by Martin Gardner in "Mathematical games," *Scientific American*, 230 (1974) 108–113.

2. *Games of the North American Indians* by S. Culin is a fascinating, detailed 850 page compendium of Native American games. First published in 1907 as part of the *Twenty-Fourth Annual Report of the Bureau of American Ethnology*, Smithsonian Institution, Washington, D.C., it was reprinted by Dover Publications, N.Y., 1975. My discussion of the game of Dish is based on descriptions found in Culin's work. Of additional interest are his many drawings of the disks and bowls used by different groups as well as references to museum collections that contain them. The similarity of the points assigned by the Iroquois and those we would predict by probabilistic analysis was also noted by E. B. Tylor in "On American lot-games, as evidence of Asiatic intercourse before the time of Columbus," *International Archives of Ethnology*, 9 (1896) 55–67. (I disagree, however, with the major point of Tylor's paper which links the Aztec game of Patolli to the Hindu game of Pachisi. For a cogent refutation of Tylor's position, see C. J. Erasmus, "Patolli, Pachisi, and the limitation of possibilities," *Southwestern Journal of Anthropology*, 6 (1950) 369–387. This paper should be read, in general, by those who believe that similarity of ideas implies diffusion.) For a discussion of Iroquoian theories of illness and dreams see A. F. C. Wallace, "Dreams and the wishes of the soul: a type of psychoanalytic theory among the seventeeth century Iroquois," *American Anthropologist*, 60 (1958) 234–248. Most recently Charles Moore has incorporated ideas about games of chance into a Native American mathematics education program. See, for example, his "Mathematics of Native American games of chance," *Kui Tatk*, Newsletter of the Native American Science Education Association, 2 (1986) 4–5. An excellent, readable introduction to probability is *Lady Luck: The Theory of Probability*, W. Weaver, Anchor Books, Doubleday & Co., New York, 1963.

3. Nim remains popular among mathematicians because the winning strategy is easily decided through the use of binary numbers. The game is discussed by M. Gardner in *The Scientific American Book of Mathematical Puzzles and Diversions*, Simon and Schuster, N.Y., 1959. According to him, the analysis and proof first appeared in "Nim, a game with a complete mathematical theory," C. L. Bouton, *Annals of Mathematics*, series 2, 3 (1901) 35–39. Gardner's book also discusses tic-tac-toe with historical references and extensions to higher dimensions.

4. Descriptions of mu torere are in A. Armstrong, *Maori Games and Hakas*, A. H. and A. W. Reed, Wellington, N. Z., 1964, pp. 9–10, 31–34; E. Best, "Notes on a peculiar game resembling draughts played by the Maori folk of New Zealand, *Man*, 17 (1917) 14–15 and *The Maori*, Polynesian Society, Wellington, N.Z., 1924, vol. 2, pp. 112–113 and *The Maori As He Was*, R. D. Owen, Wellington, N.Z., 3rd printing, 1952, pp. 148–149; A. W. Reed, *An Illustrated Encyclopedia of Maori Life*, A. H. and A. W. Reed, Wellington, N.Z., 1963, p. 57; E. Shortland, *Traditions and Superstitions of the New Zealanders*, Longman, Brown, Green, Longmans and Roberts, London, 2nd edition, 1856, pp. 158–159; and E. Tregear, *The Maori Race*, A. D. Willis, Wanganni, N.Z., 1904, p. 59. Translations of the words associated with the game were obtained from E. Tregear, *The Maori-Polynesian Comparative Dictionary*, Lyon and Blair, Wellington, N.Z., 1891 and H. W. Williams, *A Dictionary of the Maori Language*, A. R. Shearer, Wellington, N.Z., 1971. In a personal communication in 1989, Marilyn Wilkie of the Te Arawa (Maori) noted that *kawai* is most often applied to branches of family trees and that *torere* is a somewhat crude expression meaning infatuated. Since *mu* is now the word for checkers/draughts and most people she asked assumed it was a transliteration of "(your) move", she speculated that *mu torere* suggests someone crazy about such a game. The analysis of the game in this section and the sections that follow are adapted from M. Ascher, "Mu torere: an analysis of a Maori game," *Mathematics Magazine*, 60 (1987) 90–100.

5. Combinations are more usually introduced for counting the number of possible subsets of *r* items that can be formed from *n* items when the order of selection is not significant. For example, if from seven available people we wished to create a committee of four, there would be $\binom{7}{4}$ different committees possible. The committees are different in that they do not contain exactly the same combination of four people. Hence the name "seven chose four." Also, selection of a subset of *r* items from *n* can just as well be viewed as the creation of a subset of *n* − *r* items not to be selected. As a result, "*n* chose *r*" always gives the same numerical result as "*n* chose *n* − *r*."

Portions of the analysis that are only sketched here can be read in detail in the last paper cited in notes of section 4. Further, the number of circular arrangements inequivalent under rotation as an application of Pólya's theorem can be found in M. Eisen, *Elementary Combinatorial Analysis*, Gordon and Breach science Publishers, N.Y., 1969 and in R. C. Reed, "Pólya's theorem and its progeny," *Mathematics Magazine*, 60 (1987) 275–282. In general, an approach different from mine would be the use of Pólya's enumeration theory (see, for example, C. L. Liu, *Introduction to Combinatorial Mathematics*, McGraw-Hill Book Co., N.Y., 1968). Also, circular arrangements are examined by H. Perfect,

"Concerning arrangements in a circle," *Mathematical Gazette*, 40 (1956) 45–46.

6. I am indebted to Carole Frick for her assistance with Figures 4.11 and 4.13.

7. This alternate version of the rule is stated in the instructions of a commercial version of the game sold to visitors to New Zealand as an example of Maori culture. The game was purchased for me by my sister, Paula Belsey, when she visited New Zealand in 1987. It was manufactured by C. I. Witte, Christchurch, New Zealand, 1984. The same rule is found in the addendum of R. C. Bell, *Board and Table Games from Many Civilizations*, Dover Publications, Inc., N.Y., revised edition, 1979.

8. In W. Ahrens' *Mathematische Unterhaltungen und Spiele*, v. 1, B. G. Teubner, Leipzig, 1921, there is a discussion of the history of the six-person river crossing problem. Another discussion is in E. Lucas, *Recreations Mathematiques*, second ed., Gauthier-Villars et Fils, Paris, 1891. The most popular characters are the jealous husbands and their wives. The three servants and masters are found in E. Riddle, ed., *Recreations in Science and Natural Philosophy: Dr. Huttons' Translation of Montucla's Edition of Ozanam*, Nuttall and Hodgson, London, 1844; the masters and slaves in Ahrens; the merchants and robbers in N. Ausubel, ed., *A Treasury of Jewish Folklore*, Crown Publishers, N.Y., 1948; and the quarreling in-laws in H. E. Dudeney, *Modern Puzzles and How to Solve Them*, Frederick A. Stokes Co., N.Y., 1926. The missionaries and cannibals, also in Dudeney, have recently become more widespread as the problem has become popular among people studying human modes of problem solving. See, for example, S. Amarel, "On the representation of problems of reasoning about actions," in *Machine Intelligence*, vol. 3, Edinburgh University Press, Edinburgh, 1971, pp. 131–171; J. McCarthy, "Circumscription—a form of non-monotonic reasoning," *Artificial Intelligence*, 13 (1980) 27–39; and M. Hunt, *The Universe Within*, Simon & Schuster, N.Y., 1982.

Both Ahrens and Lucas (cited above) also discuss the history of the wolf/goat/cabbage problem. J. Bolte enlarges on that history in "Der Mann mit der Ziege, dem Wolf und dem Kohle," *Zeitschrift des Vereins für Volkskunde*, 13 (1903) 95–96, 311 and in another similarly titled article in the same journal twenty years later, 33 (1923) 38–39. The twelfth-century Latin hexameter is quoted from Bolte 1903, p. 95. The epistle from Alcuin to Charlemagne is discussed in F. Tupper, Jr., "Riddles of the Bede tradition," *Modern Philology*, 2 (1904) 561–572. Some mathematical recreation books in which the puzzle appears are: W. W. Rouse Ball, *Mathematical Recreations and Essays*, 10th edition, Macmillan & Co., Ltd, London, 1922; J. Degrazia, *Math Is Fun*, Gresham Press, N.Y., 1948; E. Riddle (cited above); H. E. Dudeney, *Amusements in Mathematics*, Dover Publications, N.Y., 1958; and C. Withers and S. Benet, *The American Riddle Book*, Abelard-Schuman, N.Y., 1954.

Collections of Western folklore containing the puzzle are: A. Nicolson, *Gaelic Riddles and Enigmas*, Norwood Editions, Glasgow, 1977; V. E. Hull & A. Taylor, *A Collection of Welsh Riddles*, University of California Press, Berkeley, 1942; J. Kamp, *Danske Folkeminder, Aeventyr, Folkesagen, Gaader, Rim Og Folketro, Samlede Fra Folkemende*, R. Neilsen, Odense, 1877; D. Sadovnikov, *Riddles of the Russian People*, original 1876, translated with an introduction by A. C. Bigelow, Ardis Publishers, Ann Arbor, 1986; A. Balladoro, *Folk-lore Veronese: Novelline*, Fratelli Drucker, Verona, 1900; V. Imbriani, *La Novellaja Fiorentina*, Rizzoli, Milan, 1976 (fusion of two books first published in 1871, 1872); G. Pitrè, *Fiable Novelle e Racconti Popolari Siciliani* vol. 4, Al Vespro, Palermo, 1978 (first published 1875); F. Schmitz, *Volkstümliches aus dem Siebengebirge*, Verlag von P. Hanstein, Bonn, 1901 (about Transylvania, now in Rumania); Professor Wibbe, "Der Sammler," *Niedersachsen*, 19 (1914) 503; A. Orain, *Contes du Pays Gallo*, M. Champion, Paris, 1904; A. H. Fauset, "Negro folk-tales from the South," *Journal of American Folklore*, 40 (1927) 276–292; and E. C. Parsons *Folk-lore of the Sea Islands. South Carolina*, American Folk-lore Society, N.Y., 1923.

African versions of the puzzle are from E. C. Parsons, *Folk-lore from Cape Verde Islands*, vol. 2, American Folk-lore Society, N.Y., 1923; E. Littman, *Publications of the Princeton Expedition to Abyssinia*, E. J. Brill Ltd., Leyden, 1910; S. N. Martin, *Les Devinettes du Cameroun*, Yaounde, Agracam, 1976; L. Frobenius, *Atlantis: Volksmarchen der Kabylen*, vol. 1, E. Diederichs, Jena, 1921; D. Westermann, *Die Kpelle: ein Negerstamm in Liberia*, Vandenhoeck & Ruprecht, Göttingen, 1921; and E. W. Smith & A. M. Dale, *The Ila-Speaking Peoples of Northern Rhodesia*, vol. 2, Macmillan and Co., Ltd., London, 1920. Another reference to a Kpelle version appears in J. Gay and M. Cole, *The New Mathematics and an Old Culture, A Study of Learning Among the Kpelle of Liberia*, Holt, Rinehart & Winston, N.Y., 1967. However, its statement "only two things can cross at a time" is ambiguous and there are no clarifying details or indications of the problem's source. My letter requesting further information received no response. Probably not realizing that there is another version, C. Zaslavsky's statement of the Kpelle problem (*Africa Counts*, Prindle, Weber & Schmidt, Boston, 1973) based on this reference assumes the Western logical structure. In it the characters are a leopard, a goat, and cassava leaves. If, in the 1960s, a version with the Western structure was heard from the Kpelle, it does not detract from or modify our interest in the version collected in about 1910.

For background on the Kpelle, I used, in particular, Chapter 6, pp. 197–240, "The Kpelle of Liberia" in J. L. Gibbs, Jr., *Peoples of Africa*, Holt, Rinehart and Winston, Inc., N.Y., 1965. (It should also be noted that rice, the food in the story, is the subsistence crop and main cash crop of the Kpelle.) My comments on the Swahili language follow C. W. Rechenbach, ed., *Swahili-English Dictionary*, Catholic University of America Press, Washington, D.C., 1968. Discussion of the Ila is based on both volumes 1 and 2 of the book by Smith and Dale cited above and M. A. Jaspan, *The Ila-Tonga Peoples of North-Western Rhodesia*, International African Institute, London, 1953.

I am indebted to Lori Repetti for her assistance in searching for sources of this puzzle.

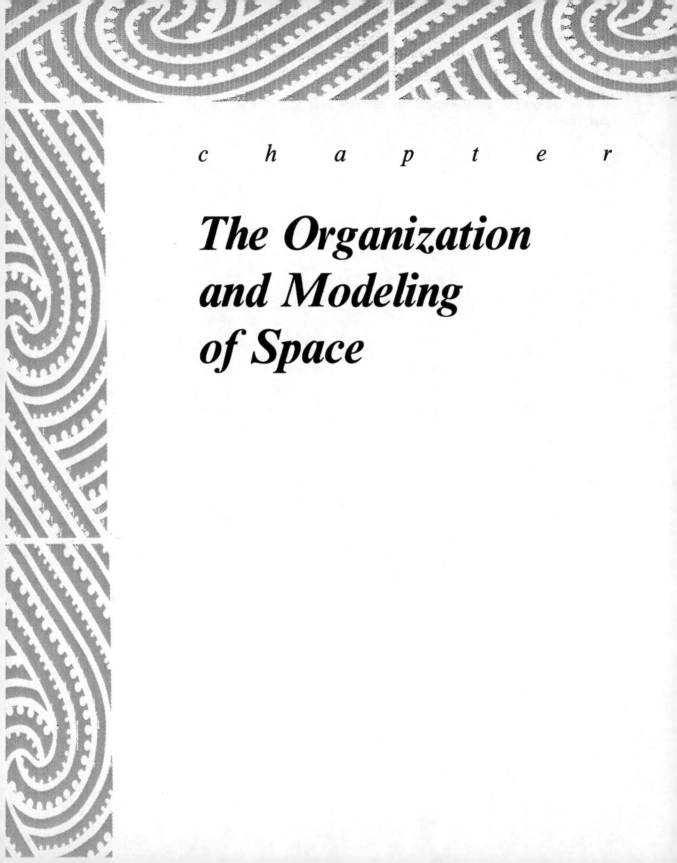

c h a p t e r

The Organization and Modeling of Space

f   i   v   e

1 If you look around the room in which you are sitting, you will probably see many straight lines, flat surfaces, and right angles. Look, for example, at a corner where the floor and walls meet, at a door frame, at a window and its frame, or at your desk. If there are any wall decorations, they too probably contain straight lines and are in rectangular frames with the bottom edges set parallel to the floor. We define the space around us by both physically and mentally imposing order upon it. More and more in our Euro-American culture, as exemplified by the enlarging and changing cityscape in addition to interior design, that order is made up of lines, rectangles, planes, and rectangular solids. To most of us these forms are necessary, sensible, and, above all, proper.

The Line

... in every human culture that we will ever discover, it is important to go from one place to another, to fetch water or dig roots. Thus human beings were forced to discover—not once, but over and over again, in each new human life—the concept of the straight line, the shortest path from here to there, the activity of going directly towards something.

In raw nature, untouched by human activity, one sees straight lines in primitive form. The blades of grass or stalks of corn stand erect, the rock falls down straight, objects along a common line of sight are located rectilinearly. But nearly all the straight lines we see around us are human artifacts put there by human labor. The ceiling meets the wall in a straight line, the doors and window-panes and table-tops are all bounded by straight lines. Out the window one sees rooftops whose gables and corners meet in straight lines, whose shingles are layered in rows and rows, all straight.

The world, so it would seem, has compelled us to create the straight line so as to optimize our activity, not only by the problem of getting from here to there as quickly and easily as possible but by other problems as well. For example, when one goes to build a house of adobe blocks, one finds quickly enough that if they are to fit together nicely, their sides must be straight. Thus the idea of a straight line is intuitively rooted in the kinesthetic and the visual imaginations. We feel in our muscles what it is to go straight toward our goal, we can see without eyes whether someone else is going straight. The interplay of these two sense intuitions gives the notion of straight line a solidity that enables us to handle it mentally as if it were a real physical object that we handle by hand.

By the time a child has grown up to become a philosopher, the concept of a straight line has become so intrinsic and fundamental a part of his thinking that he may imagine it as an Eternal Form, part of the Heavenly Host of Ideals which he recalls from before birth. Or, if his name be not Plato but Aristotle, he imagines that the straight line is an aspect of Nature, an abstraction of a common quality he has observed in the world of physical objects.

Reprinted from *The Mathematical Experience*, by Philip J. Davis and Reuben Hersh, by permission of Birkhäuser Boston Inc. Copyright 1981.

For us, what could be more natural than looking up at the sky, spotting particular stars, mentally connecting the star-points with straight line segments, and creating constellations that are seen by generation after generation? It could, of course, be otherwise. In keeping with their spatial ideas, native Andean peoples see other constellations far more irregularly shaped, made up of darker and lighter blotches (clouds of interstellar dust) in the sky. Or contrast the two boxed statements; one is a statement about the line made by two American professors of mathematics, and the other is a statement about the circle made by Black Elk, an Oglala Sioux.

While they differ on the geometric form, the writers share their degree of conviction in the rightness of their ideas and support their view with nature, God, achievement of goals, and proper human development. Taken separately or together, the statements highlight the fact that geometric ideas are an integral part of a culture's world view. Western cosmological ideas have, of necessity, influenced the course of Western

The Circle

... I am now between Wounded Knee Creek and Grass Creek. Others came too, and we made these little gray houses of logs that you see, and they are square. It is a bad way to live, for there can be no power in a square.

You have noticed that everything an Indian does is in a circle, and that is because the Power of the World always works in circles, and everything tries to be round. In the old days when we were a strong and happy people, all our power came to us from the sacred hoop of the nation, and so long as the hoop was unbroken, the people flourished. The flowering tree was the living center of the hoop, and the circle of the four quarters nourished it. The east gave peace and light, the south gave warmth, the west gave rain, and the north with its cold and mighty wind gave strength and endurance. This knowledge came to us from the outer world with our religion. Everything the Power of the World does is done in a circle. The sky is round, and I have heard that the earth is round like a ball, and so are all the stars. The wind, in its greatest power, whirls. Birds make their nests in circles, for theirs is the same religion as ours. The sun comes forth and goes down again in a circle. The moon does the same, and both are round. Even the seasons form a great circle in their changing, and always come back again to where they were. The life of a man is a circle from childhood to childhood, and so it is in everything where power moves. Our tepees were round like the nests of birds, and these were always set in a circle, the nation's hoop, a nest of many nests, where the Great Spirit meant for us to hatch our children.

But the Waischus (whitemen) have put us in these square boxes. Our power is gone and we are dying, for the power is not in us any more. You can look at our boys and see how it is with us. When we were living by the power of the circle in the way we should, boys were men at twelve or thirteen years of age. But now it takes them very much longer to mature.

Well, it is as it is. We are prisoners of war while we are waiting here. But there is another world.

mathematics, but mathematics, in turn, reinforced those ideas through art, architecture, measuring and mapping schemes, ways of seeing and describing, and even our aesthetic sense. Before we elaborate on the ideas of some other cultures, let us talk a bit more about our own ideas and their ramifications.

2 Until the late nineteenth century, Euclidean geometry was believed to both describe and reveal truths about the physical world. Because of its intimate connection with our world view, more than 2000 years elapsed before the assumptions on which it was grounded were fully clarified. Euclid's work was the model for Western mathematics; there were definitions and postulates and axioms from which theorems were logically derived. But, since Euclid's time (about 300 B.C.E.), some people have been concerned that one of the postulates was not independent of the others and so should be eliminated and, instead, proved as a theorem. It was this concern that led to the realization that there could be other postulates resulting in other theorems and, hence, geometries that were different from Euclid's. The question then became, which of these geometries was true? Or, even worse, if different postulates led to different geometries, what was the relationship between geometry and the truth about space?

Basic to Euclidean geometry are points, lines, surfaces, and solids. Between any two points, exactly one straight line can be drawn and any straight line can be continued endlessly. Given any radius and any point as a center, a circle can be drawn. (Even the circle depends on a line; a radius is a straight line segment of fixed length.) A straight line can be set upon another straight line so that adjacent angles are equal; these angles are called right angles and all right angles are equal. And, through a point not on a line, there is just one line parallel to the first (by parallel it being understood that the two lines do not intersect). All of this rests upon what is meant by points, straight lines, and flat surfaces, and on the belief that these can be used to separate space into parts. That is, for example, one part of a surface can be separated from another by a boundary line or a line can be separated into two parts by a point. Further, it is assumed that space has three dimensions, space is continuous (there are no gaps in it), space is infinite (extends without bound in all directions), space is uniform (size and shape do not change because something is in one place rather than another), and space has zero curvature everywhere.

By contrast, in one non-Euclidean geometry, Riemannian geometry, lines that are extended return upon themselves and more than one straight line can be drawn between a pair of points. In Riemannian geometry, space still has three dimensions and is continuous and uniform. However, space has positive curvature. In another, Lobochevskian geometry, space has negative curvature. Curved three-dimensional space is difficult to describe and just about impossible to imagine. The most common approach to at least partial visualization is to reduce the issue to two dimensions. A surface of a basketball is a two-dimensional surface with positive curvature that is the same everywhere. A region of the surface around any point would all be on the same side of a plane tangent to the point. (The surface of a football also has positive curvature everywhere, but the curvature varies from place to place.) The interior surface of a doughnut, on the other hand, has uniform negative curvature. For a plane tangent to any point on this, part of the surface near the point would lie on one side of the plane and part on the other side. Another way to visualize these two-dimensional surfaces and their tangent planes is to imagine trying to balance a book. Using the flattened palm of your hand or your outstretched forearm would present few problems as they are akin to surfaces of zero curvature. With care, you can balance the book on the top of your head (a surface with positive curvature), but not even the smallest book can be placed where your neck meets your shoulder because it is a saddle-like surface with negative curvature. One example of the effect of these differences in curvature is the length of the circumference of a circle. If one end of a thread of length r is held fixed and the other end is swung around (while constraining the thread to lie on the surface), the figure described by the free end has a circumference of exactly $2\pi r$ on a surface of zero curvature, a circumference greater than $2\pi r$ on a surface of positive curvature and less than $2\pi r$ on a surface of negative curvature. How much greater or less than $2\pi r$ depends on the specific curvature. For three dimensions, the descriptions involve hypersurfaces of hyperspheres and hypersurfaces of hyper-pseudospheres. For us it is sufficient to stop here, having made the point that Euclidean assumptions about space were found not to be the only ones possible.

From all of this came the understanding that each of these geometries, and there can be others, is an abstract system; their postulates and theorems have nothing to do with the *truth* about space. This changed the concept of mathematics itself. Mathematical constructs may or need not be approximations of reality. At times, however, depending on the question or situation that is being considered, different mathematical constructs may fit closely enough to be useful. As for the nature of space, that became an empirical issue and was passed on to the people called physicists rather than mathematicians. Mathematical formulations and mathematical models remain, however, the mode through which the findings of physicists are expressed.

For many cultures, as well as for our contemporary physicists, space and time are so inextricably interwoven that one cannot be discussed without the other. In Western culture at the end of the seventeenth century, primarily associated with Isaac Newton, the three spatial dimensions were augmented by a fourth dimension, namely time. This time dimension, however, was not intertwined with the other three. A configuration in space could change with *the passage* of time, but spatial properties remained absolute and were not affected by time. This idea was drastically changed by Einstein's theories of relativity in the twentieth century. Space and time became interrelated and what were previously absolutes became dependent on one's frame of reference. According to the special theory of relativity, travelers watching earth from two different spaceships, moving at different speeds, would have different opinions on, for example, the length of time it took for an apple falling from a tree to hit the ground. Their opinions would also differ on the distance the apple fell. Most crucial is that neither would be more or less correct; neither the distance nor the time of fall is absolute. The interweaving of space and time becomes even more complex with the general theory of relativity. In it, the presence of matter is also recognized as affecting geometric properties. As a result, the mathematical model is one in which space-time is curved and the curvature is irregular. To all of this we must add the big bang theory. According to it, the universe started as a point and has been expanding ever since. The expansion is not taking place *in* space but *of* space. Some cosmologists believe that the expansion will continue forever and others believe that it might slow or even reverse so that the universe returns to but a single point.

On a day-to-day basis, for most of us in Western culture, Euclidean spatial concepts still prevail. One of the reasons people believed that Euclidean geometry had to be true was that they could not visualize another. Our mathematicians and physicists have tried to expand our visualizations. Here we have a case in which the concepts of some of our wise men are not an integral part of the world view of our culture. In a sense, their ideas are as strange to most of us as if they were from another culture. Notions of space and time are so crucial to the framework we use to perceive, structure, and interpret experience that it is extraordinarily difficult to even conceive of others. Hence, as we turn to other cultures, it should be kept in mind that, as a result of the constraints of our own, other spatial-temporal ideas will be particularly hard to visualize.

The Navajo are a Native American people whose population and cultural center, since 1869, is a reservation on a high mountainous plateau that extends across northwestern New Mexico into northeastern Arizona and slightly above that into southern Utah. Despite considerable interchange and assimilation with the surrounding dominant culture, the Navajo world view remains vital and distinctive.

The Navajo believe in a dynamic universe. Rather than consisting of objects and situations, the universe is made up of processes. Central to our Western mode of thought is the idea that things are separable entities that can be subdivided into smaller discrete units. For us, things that change through time do so by going from one specific state to another specific state. While we believe time to be continuous, we often even break it into discrete units or freeze it and talk about an instant or point in time. Or, we often just ignore time. For example, when we speak of a boundary line dividing a surface into two parts or a line being divided by a point, we are describing a static situation, that is, one in which time plays no role whatever. Among the Navajo, where the focus is on process, change is everpresent; interrelationship and motion are of primary significance. These incorporate and subsume space and time.

Let us take first the idea of subdividing things into parts. We see the human body, for example, as a physical unit. We also describe it as having distinct parts such as arms, legs, a heart, etc. Although we recognize, say, blood pressure as significant, to us it is not a *part* of the body in the same sense; blood pressure has no specific location and no specific boundaries. To the Navajo, on the other hand, blood pressure is a part of the body. The idea that something is a part of the body means, first and foremost, that it is involved in making the body work. The dynamic whole that is the body is a system of interrelated parts.

Of course, for the Navajo, spatial boundaries and borders do exist. But even they are viewed as having dynamic components. There are boundaries that require action in order to go around or over them, but whatever was ongoing on the first side can continue on the other side; other boundaries require that actions be modified or reoriented. A mountain ridge within Navajo territory is a boundary of the first type, while a mountain ridge separating Navajo and non-Navajo territory is of the second type. As for the actual position of the mountain ridge, to us position is defined as where something *is*; the Navajo view it as the result of the withdrawal of motion—the mountain ridge is in the process of being in a particular place. And, continuing with a mountain ridge as an example, motion and process are further involved. The mountain ridge itself is an interrelated system of parts that are in motion and in the process of change. Further, the entire earth, of which the mountain ridge is an integral part, is in motion as well. The earth and sky are always undergoing expansion and contraction. Starting from a center, they are being stretched outward in a clockwise spiral and will eventually shrink back to the center. The motion is continuous, systematic, and gradual, and will retrace the same path during shrinkage as is being followed during expansion. To us, the significant aspect of a boundary is that it is a spatial divider; to the Navajo, the significance is the processes of which the boundary is a part and how it affects and is being affected by those processes. The Navajo react quite negatively when fences are placed upon their reservation land. One major reason is their belief that space should *not* be segmented in an arbitrary and static way.

The description of two surfaces that overlap is another specific example contrasting a Navajo concept with a Western one. To us, when we think of one surface overlapping another, what is significant is that they have a region in common. To the Navajo, the fact that two things overlap is one aspect of some active happening. Of primary importance is whether or not the same or different elements are in contact during the happening. Just as the overlap of a snake and rock is of one type if the snake is sleeping on the rock and of another if the snake is slithering over the rock, so too the overlap caused by a blanket lying on the floor must be distinguished from the overlap caused by a passing cloud's shadow on the ground. We too, of course, can make such distinctions if we wish. But—and that is a crucial *but*—to us it makes sense to think about overlap without motion, whereas to the Navajo, motion is intrinsic to overlap just as it is intrinsic to everything else.

It is from studies of the Navajo language that we have some under-standing of their view of space-time. In fact, it is mostly through language that we can understand how any culture comprehends and structures the world. Most aspects of the world are perceived, selected, organized, and reacted to in ways that are learned. And, since so much of human learning comes through language, the language of a culture is the key to its metaphysics and world view. In Navajo, verbs predominate and there are, for example, 356,200 conjugations of a verb translatable as "to go" and little, if any, use of an analogue to our verb "to be." Descriptions of action, events, objects, and even the statement of concepts, emphasize movement and contain detailed elaboration of the movement's kind and direction.

Some years ago, a group of Navajo participated in a filmmaking project. Part of the purpose of the project was to have the Navajo create films that would direct others in seeing what they consider significant. A surprising result was that walking dominated almost every film. One film was said to be about the weaving of a rug. Fully fifteen minutes of a twenty-minute film showed the weaver walking—walking to collect materials; walking to shear the wool; walking to the loom; and, always, beginning and ending the walking by leaving and returning to her hogan. The process of weaving, then, is the interrelation of many processes that emanate from a center, spread out from the center, and return back along the same path. Another film, about a silversmith, was quite similarly structured; again primarily walking and, again, always starting from and returning to the hogan. Part of the reason this was surprising is that the Navajo, in fact, do not walk that much and, in fact, do not themselves carry out some of the processes shown in the films. They explained, however, that this mode was selected because it was appropriate in order to convey what they deemed most essential.

The filmmaking project also gives us insight into the intricacy of Navajo perception and understanding of motion. As a first practice attempt at splicing and editing a film, one of the Navajo cut up and reassembled some footage of two boys on a seesaw going up and down several times. She was easily able to locate where, for three or four frames, changes in position were virtually absent. These were at the end of arcs in the seesaw's trajectory. Cutting there, she reassorted the pieces, yet replicated the rhythmic, seemingly continuous, seesaw motion. While we have learned to focus on when and where something is, the Navajo focus on whether it is arriving or leaving, speeding up or slowing down, moving aimlessly or with a goal, etc. Only with this kind of information is a description considered meaningful or complete.

One final example may help to better envisage some of these Navajo concepts. For us, although relatively few, there are also some things that have motion so intrinsic that they have no meaning without it. One of these is the wind and another is a water current. Neither is amenable to separation into spatial subunits, nor can it be pinpointed in or disassociated from space/time. As we note some of our thoughts related to wind, we can begin to form a mental model with broader applicability. Furthermore, at the same time, we will gain insight into the important role of the wind itself in the Navajo world view.

When we talk about winds, we are generally concerned, at a minimum, with their speed and direction and whether they are gusty or steady. We usually include whether they are speeding up or slowing down and where they are coming from or about how long they will remain. We know that winds differ depending on having just passed over water, flat land, or a mountain ridge and that they behave differently in different seasons. We have a whole host of special words to describe special characteristics. A cyclone, for example, is a moving system of winds rotating about a low-pressure center in a specific direction (clockwise in the Southern Hemisphere or counterclockwise in the Northern Hemisphere). More common are the U.S. Weather Bureau distinctions among breezes (light, gentle, moderate, fresh, or strong), then gales (moderate, strong, whole), and then storms and hurricanes. If you live on the coast of Maine, you get prepared when you hear about an approaching Nor'easter. In Los Angeles, the Santa Ana is famous, as are the Mistral in southern France, the Föhn in Switzerland, and the Sirocco on the northern Mediterranean coast. These winds are processual happenings, not specific objects or even specific events. We cannot study the winds by holding them fixed nor can we really say where (or when) one wind ends and another begins. No matter what the wind, we are always aware that it is a local ramification of a larger system that is undergoing change due to its interrelationship with other systems.

Extending the idea of wind from something involved with atmospheric conditions to all kinds of large or small air masses in motion, the breath within us is wind as well. In the Navajo world view, wind is of central importance; it is around us and within us and is the source of animation of all that was created. Space is continuous; space has three dimensions; space is finite in that the universe will eventually retract to the center. Above all, it is the dynamic character of space and all within it that is of primary significance to the Navajo.

4 We move now to a very different area of North America—to the Canadian Arctic above the sixtieth parallel. The region is sparsely inhabited by the Inuit who live spread throughout this area and beyond it. There are many different groups within the Inuit but, by and large, with local variations, they share the same language and the same culture.

Wind and snow, cold and ice are present much of the year. It is an environment that is traversed via water in the spring and summer and over ice and snow-covered land during the long winter. What is solid one day may be water another and what is connected on one day may be separated on another. Because periods of light on winter days are extremely short, travel takes place in the dark as well as light. Although a sound knowledge of what are waterways in the summer and the ins-and-outs of the coastline is essential, a map showing the location of land masses is insufficient for winter travel. Size and shape are in constant flux. Crucial to safe travel are detailed observations of snow contours, cracks in the ice, the quality of what is underfoot, former sled tracks, the wind, the interrelationship of all these and how they are changing. These observations depend as much on sound and feel as they do on vision. And, since any path being followed is not precisely the same as one that was followed before, even stone cairns or fixed landscape features must be recognized under variable conditions of light, distance, and direction.

Travelers to the Arctic in the late 1880s and early 1900s were quite impressed with the map drawing skill of the Inuit. Maps, drawn for each other in the snow or with pencil and paper given them by the travelers, included great detail and covered as much as 500 miles. The maps were extremely accurate in terms of relative locations and relative directions, although not necessarily so in their proportions. As well as drawing details of lakeshores, islands, mountains, watercourses, and coastlines, additional information was conveyed by numerous descriptive names. One included, for example, the place with the many washed-up tree trunks; the place where the river disappears; the place where sea-birds sojourn; the place where a knot is untied (that is, a hill that makes it necessary to uncouple a small sledge trailing from a larger one); the smoothly arched hill; and that which lies above or on the other side of where one is (that is, a ridge along the river). It was partly this map-drawing facility that led the travelers to request that the Inuit use pencil and paper to make drawings of their environment. This mode of pictorial representation was added to a long, rich tradition of incising figures on bone, on antlers, and, in particular, on ivory walrus tusks.

We will look further at the space-time ideas evidenced by those representations, but first a few more general comments and an examination of a particular facet of the Inuit language that provides some insight into spatial ideas.

The Inuit have little interest in time as an entity in and of itself or in placing bygone events in chronological sequence. Process and change are significant as part of a space-time whose time dimension focuses on the immediate past, the present, and the immediate future. What happened in the distant past is not disregarded; it is viewed as an attribute of what currently is. There is an emphasis on being as specific as possible about spatial location. Flexibility and multiplicity of perspective are important. Specific units and measurements of time or space are not. The Inuit superimpose few straight lines, right angles, and vertical planes on their environment. Their winter ice houses (*igloos*) and summer sealskin tents are circular and, although the former can have several rooms, they are interconnected circular units rather than a single unit partitioned by vertical walls.

One aspect of the Inuit language that particularly deals with space, is their system of "localizers." As we said earlier, it is not that some concepts can or cannot be expressed in any language, it is rather that the language of a culture creates and reinforces particular shared habits of thought and shared habits of observation. The Inuit localizers are used when pointing out an object, place, or event. When describing the place of events, the localizers are adverbs comparable to our words *here* or *there* in the sentences "he is sitting *there*" or "he is sitting *here*." I use the words *here* and *there* to distinguish between locations near me and those further away. Whenever such a statement is made, the Inuit habitually make far more distinctions because those distinctions are necessary in order to determine the appropriate completion of the word being used.

The Inuit, first of all, have four different suffixes that roughly correspond to our *at, from, via,* and *to*; that is, "he is jumping (*at*) there," "he is jumping (*from*) there," "he is jumping (*via*) there," or "he is jumping (*to*) there." An event takes place in time as well as space, so it must be made clear whether the location referred to remains the same throughout or is the place of the beginning, middle, or end of the event. Another choice, as in our case, is between near and further away—our *here* versus *there*. Locations further away need additional specification: up, down, inside a boundary, outside a boundary, or in the same horizontal plane with no boundary. These would be comparable to our *up* there, *down* there, *in* there, *out* there, and *over* there. The next distinction is one that is markedly different; it deals with the perspective from which all the foregoing distinctions are being made. The perspective may be that of the speaker, but it need not be. Another perspective might be that of the person being addressed, or that of the last speaker, or any other person or thing. Any one of these other perspectives is indicated in the same way so the basic distinction is between the speaker as reference point and some other reference point. ["He is jumping (*from your down*) there."]

But yet another distinction is to be made before the form of the localizer can be selected. This one has to do with the relative dimensions of the place of the event. Just as the event extends through time, the place has some extension in space, and so something about the region involved needs to be included. The region is classified into one of two categories. These categories do not readily translate into any of our own, but we can designate them by *extended* or *restricted* and get some sense of them. The classification as described here applies to the place of an event but also has other usages and so applies to objects as well. Restricted places (or objects) are those whose visible limits appear to be of about equal dimensions, such as a cubic box, a ball, a house, a folded-up blanket, or a spot on the table. Extended places (or objects) are much longer than they are wide or much longer and wider than they are thick, such as a rope, a river, a spread-out blanket, a long shelf, or a long table. But anything small is restricted regardless of its relative dimensions, as are all humans and animals. Motion, however, modifies the category that applies; a man running or a ball flying through the air, for instance, is extended. On the other hand, something in slow motion, such as a grazing caribou, remains restricted, as does a baby whether it is in motion or not. Here, again, we have an interrelationship of space and time; although the static spatial extent may be restricted, the path covered through time is not. And temporary configuration in space adds another qualification: a pair of mittens lying together is restricted, but either mitten of a spatially separated pair is extended. In a sense, the location of the pair now includes the space between. (Remember the numeral classifiers discussed in Chapter 1. There, as here, the appropriate classification is clear to the speakers of the language, but not amenable to concise summary in our terms.)

In sum, the considerations involved in picking the appropriate form of a localizer designating the place of an event are: (1) whether the event begins, ends, remains, or transverses the place; (2) whether the spatial point of reference is to be the speaker or some other; and, with regard to that point of reference, (3) whether the place is near or further away and, if away, whether it is up, down, inside a boundary, outside a boundary, or horizontally the same with no boundary involved; and, for all of these possibilities with the exception of a place inside a boundary, (4) whether the place is extended or restricted.

Table 5.1 summarizes all but the first of these considerations. Labels ① through ⑪ are given to the different roots that depend on near or away and extended or restricted. Then (P1) through (P11) are the same roots but carry a prefix that distinguishes other point of reference from speaker point of reference. To each of these 22 forms in Table 5.1 would then be added one of the four suffixes that relates the temporal dimension of the event to the place. Thus, in all, one of 88 forms is created. The system of localizers, however, does not end with these adverbs. It is also used when just pointing out a place ("Put the book *there*") or pointing out an object ("Put *this* in the box"). In grammatical terminology, the system includes predicative particles and demonstratives. Thus, "Put this there" would involve selecting the appropriate form for *this* and the appropriate form for *there*; for example, "Put this (*long thin* thing in *my here*) (on the *more or less equidimensional* thing in *your up*) there." By one count there are 686 words in the total system. Habitual use of this system of localizers clearly reinforces precision in observing and transmitting information about location. But, from it, we can also see how time and motion, as well as configuration, become a part of the clear specification of place. And, perhaps even more important, there is no presumption that perspective is fixed or shared; location is always relative, that is, a point of reference is clearly established but changeable and can be the speaker, some other person, or even some object.

Table 5.1. Considerations in the selection of a localizer

			Speaker reference point		Other reference point	
			Extended	Restricted	Extended	Restricted
Here			①	②	(P1)	(P2)
There	Same horizontal plane, no boundary		③	④	(P3)	(P4)
	Vertical disp.—up		⑤	⑥	(P5)	(P6)
	—down		⑦	⑧	(P7)	(P8)
	boundary—outside		⑨	⑩	(P9)	(P10)
	—inside		⑪		(P11)	

A visitor showed the Inuit some photographs, and because they viewed the pictures just as they came into their hands (that is, according to the visitor, upside down), he concluded that they didn't understand how to "read" a picture. Another visitor found it odd that pictures clipped from magazines were posted every which way on walls in the homes of the Inuit. He then realized that the local children found his turning of pictures before viewing them very peculiar, and in fact very humorous. To better understand what was going on, the visitor sketched about twenty differently oriented figures on a sheet of paper. The Inuit, holding the paper as handed to them, could identify each figure without any problem, but, although he had drawn the figures, the visitor had to keep turning the paper in order to see that their identification was correct. For us, there is only one way to view a picture, and we believe that way to be both necessary and correct. The way of viewing is related to the way we compose pictures and, for this too, we believe there is only one way that is realistic. Points, straight lines, right angles, and our Euclidean world view in general, play a large role in this idea of reality. There is, first of all, a flat two-dimensional rectangular surface that we assume to be perpendicular to the ground, and it has a *bottom* edge which, when we hold or hang the rectangle, will be parallel to the floor. Everything in the picture shares this same orientation.

Many of our pictorial conventions began in Europe in the fifteenth century. They were influenced by, and reinforced, particular mathematical ideas. The study of optics, going back to Euclid in 300 B.C.E., depended on rays traveling in straight lines between the eye and objects being viewed. The fifteenth-century concern was where those lines would intersect a perpendicular plane placed between what was being viewed and the viewer. Time became frozen, and hence the issue of time and the issue of motion were eliminated. Not only was time frozen in what was being viewed, but the position of the viewer was also frozen. The entire picture could only show what a viewer could see from a single fixed place in a single fixed instant; one could not show simultaneously what could be seen from above, below, behind, and inside as well as outside. Related to the fixed vantage point in front of the picture plane are the horizon line and "vanishing points" of a picture. These are where sets of parallel lines would *appear* to us to meet. Based on them are the sizes of objects that are to be seen as being at different distances from us. To show recession, the objects diminish proportionally. Westerners, whether they be creators or viewers of pictures, are taught these conventions. Others are used by other peoples. Clearly the Inuit do not share our conventions. For them, time and space remain unified and the contents of the picture is not confined to what can be seen from a single fixed position in space-time. An event through time can be depicted by showing its interrelated parts, each in its most significant aspect, despite the fact that they could not necessarily be seen at the same time nor from a single place. In general, a picture has no background and no borders; whatever is shown stands out in a featureless space that extends indefinitely in all directions. What is most distinctive is that different parts of a picture can have different ground lines and, for any part, there can be multiple vantage points. As a result, a picture need not be oriented in any particular way. The vantage points, however, are sufficiently far away that horizon lines and vanishing points, even if there were any, would be at infinity. This, combined with the lack of background, means that there are no changes in size due to differing distances; everything in the picture is relatively at the same distance.

Three Inuit drawings done in the 1920s are included here to show some of these features. Figure 5.1 shows a song festival. These festivals are held during the dark winter. The several families living in the same village each prepare by careful practice in their own homes. The songs are presented when the community gathers in a well-lit, specially built hut or one of the larger homes in the village. The leader in each song is accompanied by a chorus grouped around him. While singing and beating upon a hand-held drum, the leader stands in place but moves with rhythmic swaying motions combined with rising and sinking motions. In this picture, we can see inside as well as outside the hut, but because it has a single groundline, we would probably have no question about how *we* would hold it or view it.

Figure 5.1. Song festival in a snow hut. Drawn by Alorneq.

Figure 5.2 is a musk-ox hunt. Although the Inuit would not do so, the reader is encouraged to turn the book and view the picture from many different directions. Or, better still, look carefully at the picture without turning it or your head and notice how you are drawn into the ongoing activity.

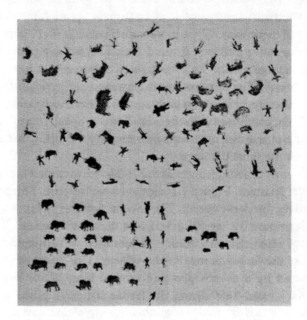

Figure 5.2. Musk-ox hunting on North Somerset Island. Drawn by Itqilik.

Figure 5.3 is rich in visual detail as well as in information about the fishing techniques of some Inuit groups. The men with hook-spears are assembled around a stone dam in a river a short way from the river's outlet into a lake. The first fish of the season are caught around the beginning of June, when they run from the lakes to the sea. This picture shows early autumn as the fish, now fatter and more valuable, return upriver from the sea (→ shows the direction of the fish, ← shows the direction of water flow). Also, in early autumn, after the snow has begun to fall but the ice is not yet very thick, another mode of fishing is used. A small snow hut is built on the ice. The snow is removed from small footprint-sized spots around the hut as well as from over a semicircle behind it. Through these "windows" in the ice, light passes into the water attracting the fish. A domed catch basin is cut into the ice floor of the hut. The oblong basin, wider at the top and narrowing to a thin slit at bottom, is capped by a dome of snow with a small opening. Through the opening, the fisherman watches and spears the fish as they enter the catch basin. The construction of the catch basin, the hut, the holes and semi-circle for light, the fisherman, and his spear are all visible in the drawing. Although both relate to autumn fishing, there is no direct relationship between the time and place of the dam and the time and place of the snow hut. As well as multiple vantage points, there are multiple ground lines; we see from everywhere and from nowhere in particular. The flow of the river was used to position the figure on the book page, but any number of different orientations are equally plausible.

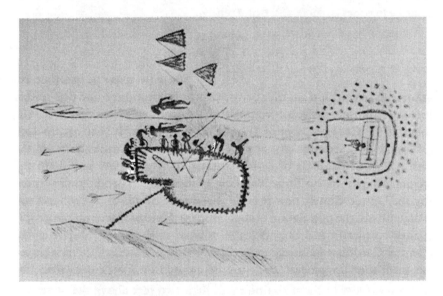

Figure 5.3. Catching fish in a *saputit* and in a *kapisilingniarfit.* Drawn by Qavdlunâq.

In the past seventy years, as is the case worldwide, Western products, Western education, and Western communications media have been superimposed on the Inuit culture. Their graphic tradition continues, but in more recent drawings, some of the characteristic features are combined with or have given way to those more common to Western culture. Nevertheless, the tradition remains quite distinctive because of the underlying difference in ideas about space-time.

When discussing the Navajo space-time concepts, we focused on wind as a means of deeper insight. It has been suggested that the Inuit concept of space-time can best be understood by realizing that it is auditory as well as visual. Sound is all around us; we need not orient ourselves or, for example, a radio in any particular way in order to hear what is being said, and sound is not stopped or defined by static structural enclosures. While we think and speak of space in primarily visual terms, the Inuit do not. Motion is intrinsic to Inuit space-time, but in a way quite different from the Navajo. Space-time is continuous, infinite, and not to be fragmented. Shape and size within it are transient. Reference points for location have to be continually established since reference points themselves are not presumed to be fixed. Any object or occurrence defines its own space, but with great specificity, that space can be located relative to other objects and occurrences.

5 We now move to another region of the world and another culture whose environment and day-to-day life are strikingly different from that of the Inuit and from our own. The place is the Central Caroline Islands in Micronesia. It is an archipelago of islands that extends east and west for about 1500 miles between the 5th and 10th parallels of north latitude. Between 1947 and 1986 the region was part of the Trust Territory of the Pacific Islands administered by the United States; now it is the Federated States of Micronesia and Palau. In all, the population is about 50,000. Here we focus on the people linked culturally and linguistically in what is shown on Map 4 as the Central Carolines language area. The islands and atolls they inhabit are so small that, in general, each can be crossed on foot in less than two hours and most of them rise no more than two feet above the water all around. (Atolls are coral islands consisting of reefs surrounding lagoons.)

In this environment, water is the predominant feature. Paddling canoes and sailboats are the mode of transportation. Open-sea fishing as well as fishing and collecting in the reefs and lagoons of the atolls are the major food sources. But the water provides more than food; it is also the means of communication. Although the Caroline Islands are far apart, they are not isolated from each other. There is continual linkage and exchange via sea travel. Not only are material resources exchanged, but there is ongoing interaction due to kinship and religious, political, and social ties. Storms and illness can be devastating to such low-lying areas

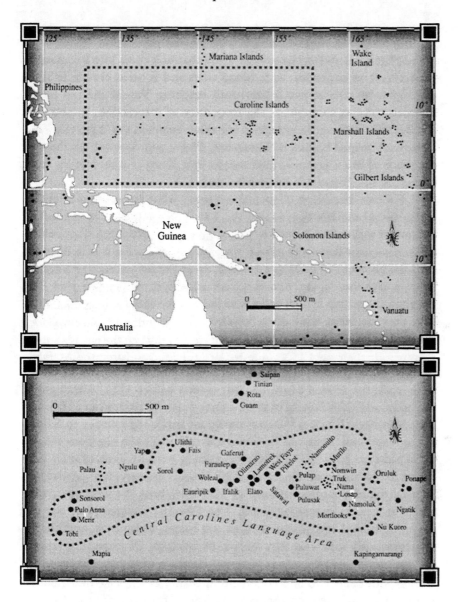

Map 4. The Caroline Islands in the North Pacific

with such small populations. Assistance in emergencies even includes sending people to help reestablish a decimated population. The islands have obligations and duties to each other that link them into a somewhat hierarchical arrangement: each island owes and is owed certain tributes and rights by one or two others until reaching Yap at the top of the hierarchy. As an example of the frequency of interaction, in one year (1962–1963) fifty-seven voyages were made between the three close and closely connected islands of Lamotrek, Elato, and Satawal. In the vast sea space of the Carolines, *close* means that Elato is fifteen miles from Lamotrek and Satawal is forty miles away.

For our discussion of spatial concepts, what attracts our attention is the spatial models that enable the Caroline navigators to sail safely, well out of sight of land, even to places they have never been. Particularly if you have ever done any sailing, you can imagine what it is to be at sea with the problem of making a landfall on an island perhaps two square miles in area that is, say, two hundred miles from your home port. The Caroline Island navigators are world famous for their navigational skill; how they do what they do is viewed with wonder by accomplished Western sailors. In addition to frequent travel between the islands spread throughout their own 1500 mile region, they have visited the Marshall Islands, New Guinea, and the Philippines. Annual journeys to Guam were made for several centuries, but ceased before 1910. Although no one had made a similar trip in at least sixty years, in 1969, in an unfamiliar European-style ketch, a Puluwat navigator made the journey to Saipan in the Marianas. That trip involved 450 miles of open sea from Pikelot to Saipan (see Map 4). And, to dramatically present some of their land claims to the high commissioner of the Trust Territory, navigators from Satawal made a similar journey in their own boat the next year, with others following in ensuing years. The Caroline navigators do not use any navigational equipment such as our rulers, compasses, and charts; they travel only with what they carry in their minds. Of central importance are two spatial models; one is an approximation of reality and the other they do not believe to be a statement of reality at all. Like most mathematical models, they are abstractions from which relations and logical implications can be derived, created to deal with all possible journeys rather than only journeys already taken. The models are used to organize fixed data, incorporate realistic cues, and make decisions accordingly. Their use is standardized throughout the region and is unique to the region. The navigators know more than just these models, but the models are crucial. Given the complexity of the navigators' knowledge and skills, it should come as no surprise that Westerners are still not clear on exactly how the mental spatial models enable their results.

Being a navigator is the most prestigious occupation that one can have. It requires long and intensive training, much of which takes place on land. Knowledge is considered to be property, so it is not given away freely; it is passed from father to son or from mother's brother to sister's son or to another only if a sufficiently large sum is paid. As a group, the navigators guard their secrets, but once one is an accepted member of the profession, information is freely shared. Cultural taboos reinforce the special, privileged nature of the group and, at the same time, facilitate their sharing of information. Navigators, for example, can only eat food that is separately prepared and separately served; no matter where they eat, they eat only with each other.

Fundamental to the knowledge of a navigator is a star compass based on the rising and setting position of certain stars with reference to the circle defined by the horizon. The sky is divided by the compass points into a series of latitudinal paths, and all stars that rise or set at the same place are considered to follow the same path. An underlying assumption is that the east-west axis of their islands is both the terrestrial and celestial equator. Their model is a fair approximation of reality because they are close to the equator and their region is latitudinally confined. The star compass is shown in Figure 5.4. It has 32 points, not evenly spaced, with more of them in the more frequently traveled east-west direction. Use of the model does not depend on the visibility of these stars during a journey. Its fixed points provide a framework for all star knowledge. Any stars, anywhere in their trajectories across the heavens, that can be seen even briefly during the night can be used to orient the entire sky.

The compass points are the basis for describing direction. To sail from one place to another is to sail, for example, to the setting of star A with a wind coming from the rising of B. The places of the many inhabited and uninhabited islands, reefs, sandbanks, and numerous other significant sea features are learned with reference to this system. Traditional metaphors and mental imagery are used in the memorization process. The apprentice is told, for example, to imagine that he is extending a breadfruit picker from one island to another and then drawing it back or that he is pursuing a fish that keeps eluding him. Each of these is accompanied by sets of specific places and the compass directions that connect them. For example, the course of a certain fish is given as:

A to B under star X,

B to C under star Y,

C to B under star Z,

B to D under star W,

> *D* to *B* under star *M*,
>
> *B* to *E* under star *N*, and so on.

In learning and repeatedly using these images, the navigator mentally crisscrosses the sea following paths that are directed line segments between pairs of points. It must be kept in mind, however, that the images are dynamic and from the point of view of the navigator—they are not our bird's-eye view when looking at or drawing lines upon a static map.

To know the relative locations of islands and the directions to sail from one to another is a large part of navigational knowledge. But there is another, even more challenging problem. The problem is to know, while you are sailing, exactly where you are. When we walk on city streets, we can conclude by landmarks, such as buildings and street signs, how far we have come and how far we have left to go. We can easily verify that our intended left turn is at the next corner. When not following a road, there is, first of all, the problem of knowing the direction in which you

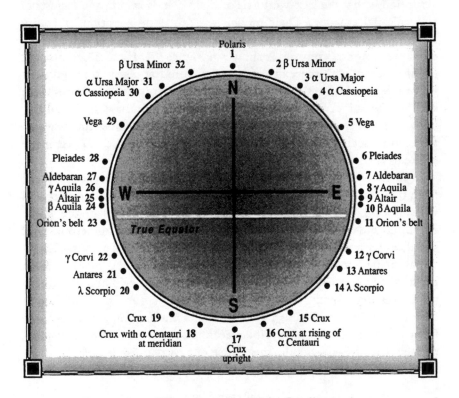

Figure 5.4. The star compass of the Caroline navigators

are heading and then, somehow, keeping track of the distance that has been covered in that direction. And, if different directions have been followed for different distances, these have to be combined to determine an end point. In sailing, the problem is compounded by the fact that it is not that easy to keep a straight heading. To go in a desired direction, the boat's course must be set in a somewhat different direction, determined by the direction and strength of the wind, the direction and strength of the current, and the angle of the setting of the sail. If, for example, the desired direction of travel in Figure 5.5 is *A* and the wind is coming from *B*, the boat must travel in some intermediate direction *C*, because it will be pushed sideways as it goes forward. The sidewards slippage away from the wind is, in our terms, the leeway (the leeward side is away from the wind; the windward side is toward the wind). A sailboat moves forward because of the force of the wind on the sail. Depending on the set of the sail, some of the force is exerted in the direction the boat is headed and some perpendicular to that. A sailboat is designed to move forward easily but resist being pushed to the side. Nevertheless, there is always some sideways motion (see Figure 5.6). Selecting a heading is, essentially, a solution to an implicit vector problem. The Caroline sailing boats resist the sidewards push better than do European boats because the hulls are asymmetrical with extensions on both sides; an outrigger to one side and balance platform on the other. The asymmetry, however, means that the sail can only be put out on one side (the platform side), compared to European boats in which it can be put out to either side. This, too, affects the relationship of the true course to the desired course. A sailboat cannot sail directly into the wind. In fact, it cannot sail in a direction too close to that either. For our boats,

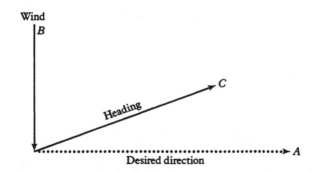

Figure 5.5. A sailboat must head in a direction somewhat different from its desired direction.

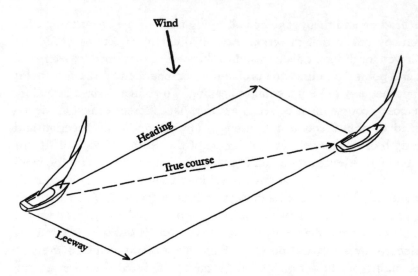

Figure 5.6.   A sailboat's course

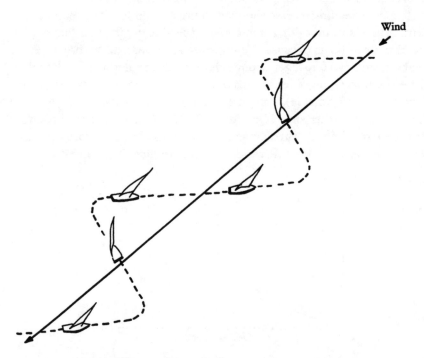

Figure 5.7.   Tacking in order to sail in the direction
of the wind

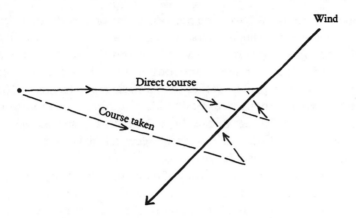

Figure 5.8. The course taken by a Caroline navigator

too close is within about 45° on either side of the wind; for the Caroline Islanders', it is within about 70°. In order to sail to points in those directions, it is necessary to *tack*; that is, to sail back and forth across the wind *changing the side* on which the sail is out each time the direction is changed (see Figure 5.7). To accomplish this on a Caroline Island boat, the mast and sail must be picked up and moved to the other end of the boat. What was the front becomes the back and vice versa. Thus, they generally take few tacks, but long ones, which means they must deviate far from their desired course (see Figure 5.8). Their tacks, however, get shorter and shorter as they near their goal and home in on a destination. In all, even without unexpected exigencies, the true course of a sailboat is *always* a combination of large and small deviations from the direct course.

For us, when we are not using electronic instrumentation and when there is no recourse to navigational aids such as buoys and lighthouses, *dead reckoning* (deducing where you are when you are sailing) depends on overlaying directions and distances on charts of the sea. The charts, magnetic compasses, rulers, protractors, and pencils are minimal equipment. We are, in effect, drawing a scale model of the journey on a two-dimensional surface. In order to know where we are, we recreate the path that has been taken. *Celestial navigation*, on the other hand, directly locates the spot at which the calculation is taking place. It requires accurate measurements of the angles of elevation above the horizon for two celestial bodies. These are combined with tabulated data and with trigonometric formulas to give information that enables two lines to be drawn on a chart. The position of the boat is the point on the chart where the two lines intersect. The spatial model used by the Caroline navigators involves both path of travel and the positions of stars, but it is vastly different from both of these. It is, for one thing, a continuous dynamic process rather than a static representation. With it, the navigator continuously maintains his bearings in relation to other known features of the sea. But perhaps the most unusual aspect is that the model is unrealistic—the point of view taken is that the boat is standing still and the islands are moving.

In this model, for every journey of more than about fifty miles, there is a starting place, an ending place, and a specific reference island. The reference island is off to the side, well out of sight of the course. (To get some idea of the placement of the reference island, look at Map 4: for the journey from Woleai to Olimarao, the reference island is Faraulep; and for Pikelot to Saipan, the reference island is Gaferut.) The boat is conceived as stationary and all else moves around it; that is, the sea and everything in it moves as a unit so that all interrelationships are retained. The direct course of the journey passes the stationary boat as the reference island passes under a specified sequence of stars. The journey in Figure 5.9 is in six stages, each stage being measured by the reference island having moved back by one star. At the start, the reference island is under star 1. When it has moved back so that it is under star 2, the destination has moved one stage closer to the boat. When the reference island is under star 6, there is just one stage remaining until the boat and

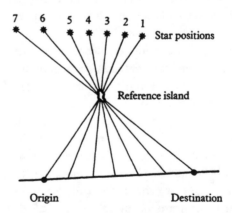

Figure 5.9. As the reference island moves under consecutive star positions, the destination moves toward the boat.

destination are in the same place. Somehow, by focusing on the motion of the reference island and its relationship to him, the Caroline navigator is able to incorporate his true course, with all its deviations from the direct course, into the model and to know where he is throughout the journey. If the direct course itself were followed, it would be difficult to understand how this is done; for the true course, it is even harder to understand.

As the end of a journey nears, the navigators use a most interesting, although not mathematical, technique to compensate for any errors due to approximation. The points in the star compass are, in a few places, as much as 15° apart. Thus, there is a limit to the accuracy of any direction given; there might be an error of as much as 5°–7°. On, say, a 200 mile journey, that alone could mean missing the island by about 25 miles (200 sin 7°). When there is just one stage remaining, the navigator keeps a sharp lookout for seabirds. If it is night, he stops and waits for daylight. Birds live on islands, but range as much as 20 miles out over the sea to look for food during the day. Once a bird is sighted, the navigator waits until dusk when he obtains clear directions from the birds as they return to their island.

For the Caroline navigators, orientation in space means orientation on the sea. Their star compass is an abstraction that enables them to define direction and, with the aid of the moving stars in the true sky, they keep the star compass oriented around them. They, as we, emphasize points and lines, but their lines are directed paths viewed from the perspective of moving along on the path (or being on a path that moves toward them) rather than visualizing its completed form statically or from above. Motion, relative position, and relative direction are of primary importance. The concern of the Caroline navigators is practical, but their modeling approach is abstract.

6 In this chapter we have ranged over several different cultures and several different expressions of spatial ideas. Space is not something easily summarized or easily described, in part because our relationship to it is so intimate and so multifaceted. We are *in* space, we move *through* space, and we act *on* space. We construct partitions of space and, using various materials, structure and organize the space around us. But we also conceptually structure space in order to have a context for perceiving and describing objects and motion within it. In many cases, the conceptual structure is not of space alone but rather of space-time, the two being so interrelated that the conception of one involves the other. Different descriptive modes are created to describe where in space (or space-time) objects and events occur relative to each other. Depending on what is important to them, each culture establishes its own spatial/temporal conventions. And for the people of each culture, their own physical and conceptual structuring of space-time is such an integral part of their world and their world view that it seems both obvious and natural.

Notes

1. The constellations of the native Andeans are discussed in G. Urton, *At the Crossroads of the Earth and the Sky: An Andean Cosmology*, University of Texas Press, Austin, 1981. I acknowledge, with appreciation, the permission granted by the University of Nebraska Press to reprint the statement about the circle from *Black Elk Speaks*, N. G. Neihardt, 1961, pp. 198–200, and the permission granted by Birkhäuser Boston Inc. to reprint the statement about the line from *The Mathematical Experience*, P. J. Davis and R. Hersh, 1981, pp. 158–159.

2. Suggested readings regarding our changing views about geometry and space-time are "On the origin and significance of geometrical axioms," H. von Helmholtz, 1870, and "The postulates of the science of space," W. K. Clifford, 1873, as reprinted in vol. 1 of *The World of Mathematics*, J. R. Newman, ed., Simon & Schuster, New York, 1956 (pp. 647–670 and pp. 552–567, respectively); "The meaning of mathematics," M. Kline, 1960, and "Non-Euclidean geometry," S. Barker, 1964 as reprinted in vol. 2 of *People, Problems, Results*, D. M. Campbell and J. C. Higgins, eds., Wadsworth Inc., Belmont, California, 1984 (pp. 11–18 and pp. 112–127, respectively). Particularly recommended is R. v. B. Rucker's *Geometry, Relativity and the Fourth Dimension*, Dover Publications, Inc., New York, 1977.

3. *The Anthropology of Space*, R. Pinxten, I. van Dooren, and F. Harvey, University of Pennsylvania Press, Philadelphia, 1983 is a detailed exploration of Navajo conceptualizations of space combining study of their language and natural philosophy. In addition, it proposes a Universal Frame of Reference to be used in any culture. The universality of this frame of reference is open to question, but, in any case, it provides a mode for studying spatial concepts. The book is highly recommended. The estimate of the number of conjunctions of the verb translated as "to go" is from G. Witherspoon's *Language and Art in the Navajo Universe*, University of Michigan Press, Ann Arbor, 1977. This book and his citations from the work of Gladys Reichard provide important insights as do M. Astrov, "The concept of motion as the psychological leifmotif of Navaho life and literature," *Journal of American Folklore*, 63 (1950) 45–56, and H. Hoijer,

"Cultural implications of some Navaho linguistics categories," *Language*, 27 (1951) 111–120. The latter follows the approach of B. L. Whorf, who focused on the Hopi. In particular, Whorf's contrast between Hopi and Western metaphysics is in "An American Indian model of the universe," *International Journal of American Linguistics*, 161 (1950) 72. The filmmaking project is discussed in S. Worth and J. Adair, "Navajo filmmakers," *American Anthropologist*, 72 (1970) 9–34 and expanded in their *Through Navajo Eyes*, Indiana University Press, Bloomington, 1972. Of general interest is E. T. Hall's *The Hidden Dimension*, Anchor Books, Doubleday & Co., Inc., N.Y., 1969 about the perception and use of space as a cultural elaboration.

4. The space-time concepts of the Inuit are discussed by E. S. Carpenter in several writings including "Eskimo space concepts," *Explorations*, 5 (1955) 131–145; *Eskimo* (with F. Varley and R. Flaherty) University of Toronto Press, Toronto, 1959; *Man and Art in the Arctic*, Museum of the Plains Indian, Browning, Montana, 1964; "The timeless present in the mythology of the Aivilik Eskimos," in *Eskimo of the Canadian Arctic*, V. F. Valentine and F. G. Vallee eds., McClelland and Stewart Ltd., Toronto, 1968, pp. 39–42; and *Eskimo Realities*, Rinehart and Winston, New York, 1973. The specifics of the system of localizers in the Inuit language are from R. C. Gagné, "Spatial concepts in the Eskimo language," pp. 30–38 in *Eskimo of the Canadian Arctic* just cited, and J. P. Denny's "Locating the universals in lexical systems for spatial deixis," *Papers from the Parasession on the Lexicon*, Chicago Linguistic Society, 1979, and his more extensive paper "Semantics of the Inuktitut (Eskimo) spatial deictics," *International Journal of American Linguistics*, 48 (1982) 359–384.

Important early investigations and reports on Inuit culture are F. Boas, "The central Eskimo," *Sixth Annual Report of the Bureau of American Ethnology, 1884–5*, Smithsonian Institution, Washington, D.C., 1888, pp. 399–675 and K. Rasmussen, *Report of the Fifth Thule Expedition 1921–24*. The parts of the latter used here were volume 7, nos. 1 and 3 and volume 8, nos. 1 and 2. These were published by Gyldendalske Boghandel, Nordisk Forlag, Copenhagen, 1929–1931. Volume 8, nos. 1 and 2 have been reprinted by AMS Press, Inc., New York, 1976. Both Boas and Rasmussen collected maps and drawings which are included in their reports. Several of these drawings, as well as several collected by Carpenter, are also reproduced in the books by Carpenter. My Figures 5.1, 5.2, and 5.3 are from Rasmussen's reports facing p. 298 and p. 247 in vol. 8, no. 1 and p. 524 in vol. 8, no. 2, respectively. I acknowledge with appreciation the permission granted by Gyldendalske Boghandel to reproduce these figures.

The continuing Inuit graphic tradition is discussed and illustrated in J. M. Vastokas, "Continuities in Eskimo graphic style," *Artscanada*, 28 (1972) 68–83; *The Inuit Print*, National Museum of Man, Ottawa, 1977, and *Keeveeok. Awake!*, Boreal Institute for Northern Studies, University of Alberta, Edmonton, 1986.

It is Carpenter who describes Inuit space-time as acoustic as well as visual.

Also, he is the visitor who reported the children's reaction to his picture turning. The earlier visitor who thought the Inuit could not comprehend a photograph was R. Flaherty who filmed *Nanook of the North* in the 1920s. His comment and reaction are quoted in "Nanook and the North," P. Rotha with B. Wright, *Studies in Visual Communication*, 6 (1980) 33–60.

For discussions of the growth of our Western pictorial conventions and their relationships to our mathematical ideas, see J. White, *The Birth and Rebirth of Pictorial Space*, Faber and Faber, London, 1972, chapters 8 and 17; J. V. Field, "Perspective and the mathematicians: Alberti to Desargues," in *Mathematics from Manuscript to Print 1300–1600*, edited by C. Hay, Oxford University Press, London, 1988, pp. 236–263; C. Geertz, pp. 102–109 of "Art as a cultural system," in *Local Knowledge*, Basic Books, Inc., New York, 1983; and particularly recommended is M. A. Hagen, *Varieties of Realism: Geometries of Representational Art*, Cambridge University Press, New York, 1986, which discusses other cultures as well. It is also noteworthy that innovative, early twentieth-century Western artists began experimenting with multiple perspective after seeing the artwork of traditional peoples.

5. The Caroline star compasses and their use are described by W. H. Goodenough in "Native astronomy in Micronesia: A rudimentary science," *The Scientific Monthly*, 73 (1951) 105–110 and in *Native Astronomy in the Central Carolines*, The University Museum, University of Pennsylvania, Philadelphia, 1953. (Figure 5.4 is adapted from his Figure 2, 1953). In the latter, he also discusses their calendar. The journeys to Saipan are described in detail in D. Lewis, "A return voyage between Puluwat and Saipan using Micronesian navigational techniques" and M. McCoy, "A renaissance in Carolinian-Marianas voyaging." These are pp. 15 and 8, respectively, in *Pacific Navigation and Voyaging*, compiled by B. R. Finney, The Polynesian Society, Wellington, N.Z., 1976. Also included as chapter 7 in this collection is S. H. Riesenberg's "The organisation of navigational knowledge on Puluwat" in which mnemonic devices and learning schemes are described. These are also discussed in "Speculations on Puluwatese mnemonic structure," P. Hage, *Oceania*, 49 (1978) 81–95. In order to view navigation in its cultural context, W. H. Alkire's *Lamotrek Atoll and Inter-Island Socioeconomic Ties*, University of Illinois Press, Urbana, 1965 and his "Systems of measurement on Woleai Atoll, Caroline Islands," *Anthropos*, 65 (1970) 1–73 are particularly useful. The latter also discusses measurement unrelated to navigation. The most comprehensive study of the Caroline navigators is *East Is a Big Bird—Navigation and Logic on Puluwat Atoll*, T. Gladwin, Harvard University Press, Cambridge, 1970. (Figure 5.8 is adapted from his figure on p. 194.) The book is fascinating reading. Cognitive psychologists have been interested by the navigators' skills and, in particular, by their mental models. Writings by them are: K. G. Oatley, "Mental maps for navigation," *New Scientist*, 64 (1974) 863–866 and "Inference, navigation, and cognitive maps," in *Thinking-Readings in Cognitive Science*, P. N. Johnson-Laird & P. C. Watson, eds. Cambridge University Press, New York, 1977, pp. 537–548; and E. Hutchins, "Understanding Micronesian navigation," in *Mental Models*, D. Gentner & A. L. Stevens, eds., Lawrence Erlbaum Associates, Hillsdale, New Jersey, 1983, pp. 191–225.

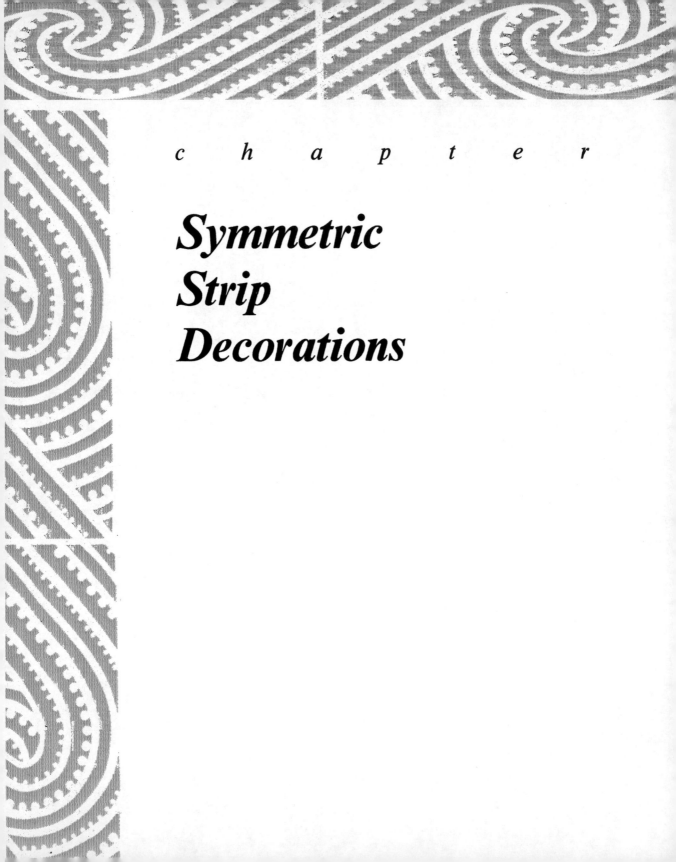

c h a p t e r

Symmetric Strip Decorations

s   i   x

The systematic organization of shapes and colors on surfaces is another way that peoples around the world have imposed spatial ordering. The creation and perception of pattern has a logical as well as aesthetic component. A mosaic floor, for example, is no more or less useful as a floor if it is unadorned, haphazardly decorated, or if it has an intricately structured pattern. For whatever reason, people expend much thought and effort on the creation of mosaics that display complex organizations of shapes and colors. Similarly, clothing, walls, eating ware, cooking ware, containers, all kinds of common artifacts, and even surfaces of the human body are frequently decorated in a patterned way.

We will focus on a particular spatial configuration which is widespread throughout many cultures. Attention will center on just two cultures, both familiar from earlier chapters, namely, the Inca (Chapter 1) and the Maori (Chapter 4). The configuration of interest here is a band or strip on which there is a single motif repeated over and over again. As we shall see, although the strips of both the Inca and Maori can be discussed with the same Western terminology, they are remarkably different, as each is expressive of and contributive to the cultural complex in which it appears. If you begin to look for strip patterns in the world around you, you can see them bordering dishes, curtains, skirts, shirts, or posters, and serving as hair bands, guitar straps, watchbands, and even gift wrapping ribbons. The example in Figure 6.1 is from one of my bathroom towels. Many of the objects you see are machine-made, so the formal repetition of a motif is a mechanical process; nonetheless, the design and use of such motifs express human choices. Whether in our culture or in another, every strip pattern has a context; that is, for example, what it is adorning, what function or meaning the object has, who created it, how it was executed, the meaning the motif has, or how it fits with other cultural expressions.

Figure 6.1. A strip pattern on a towel

When we examine repeated motifs forming strip patterns, we find that they are amenable to a formal mathematical description. The mode of description is Western; we use it here because it is a handy tool for analyzing and gaining insight into the internal logic of a strip's design. For the Malekula sand tracings, we knew the actual tracing procedures; we also observed symmetry in the results. What we observed complemented but could not replace or reconstruct knowledge of the procedures. And, from the Inuit, we know that there is more than one way to see pictorial representations. In our Western analysis of strip patterns, we make observations about symmetry and, on that basis, define categories. We cannot presume that the categories have any direct counterparts in another culture. The analysis, however, broadens our own visual vocabulary and enlarges what we are able to see. Then, as we scrutinize the patterns, we are more aware of their internal spatial relationships and, thus, more aware of the detailed spatial planning and visualization required for their execution. So before discussing the Inca and Maori strip patterns, we will look further at this Western mode of description and classification.

For different figures, there are different motions that could move the figures about but leave them apparently unchanged. Which motions or combinations of motions do this will be the basis for our categorization. Because we are focusing on figures that are constrained to remain on a strip and are repeated along it with no change in size or shape, there are only a limited number of motions possible. The motions are called *isometries*, meaning that all distances measured on the repetitions of the figure remain the same.

If, for example, a figure such as **Y** were repeated along a strip, the result would be **YYYY**. Notice that the same strip would result if the repeated motif were reflected across its central *vertical* axis as well as being moved forward. This, of course, is not true of all motifs. For example, it is not true of the motif and strip in Figure 6.2b. For that one, however, what is indistinguishable is whether the repeated motif was simply moved forward or whether it was reflected across its central *horizontal* axis as well. These motions are examples of isometries. In all, there are only four types of isometric motions possible. The first, called a *translation*, simply slides the figure forward. Another is the *reflection* of the figure across a vertical axis (as was done in 6.2a) or across a horizontal axis (as for Figure 6.2b). *Rotation* around a fixed point through 180° is also isometric. In Figure 6.2c, the point of rotation is shown on the repeated motif. And, a last type, called *glide reflect*, is a translation forward along the strip and a reflection of the figure across a line parallel to the direction of translation. In a horizontal reflection, the parts of a motif that are mirror images are one above the other, but in a glide reflect they are offset (see, for example, Figure 6.2d). These various motions can be applied separately or in combination. Altogether, however, they result in only a limited number of different categories of strip patterns, because many of the motions, when combined, form strips that are indistinguishable from each other.

In mathematical terminology, the different categories that result are called *symmetry groups*. Colloquially, the word *symmetry* evokes sameness with respect to a central vertical axis or sometimes, but less frequently, sameness with respect to a horizontal axis. In mathematics, however, symmetry has a broader meaning; the symmetries of a pattern refer to

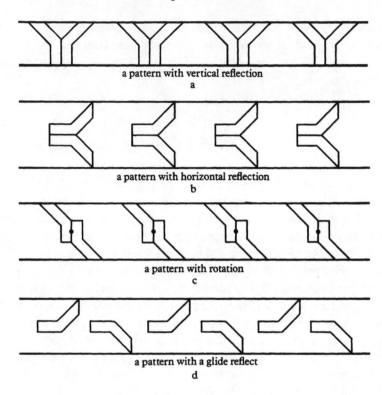

a pattern with vertical reflection
a

a pattern with horizontal reflection
b

a pattern with rotation
c

a pattern with a glide reflect
d

Figure 6.2. Repeating motifs

any isometries for which the pattern appears unchanged. Furthermore, for any strip pattern, its symmetries fall into one of seven groups; that is, for all strip patterns there are seven (and only seven) distinctly different sets of isometries that leave strips unchanged, and each of the seven sets of isometries has the mathematical properties that characterize a group. (In Chapter 3, the properties of a group were discussed. We also discussed the symmetry group of the square. That group contained eight elements. Each element was associated with a motion or combination of motions that resulted in a different orientation of the square. For any of the motions, however, the square always appeared unchanged. Here, too, the elements of the groups are associated with motions and, for a given motif, any of the motions in a single group result in a strip pattern that appears the same.) To solidify these ideas, look at the seven different strips of Figure 6.3. Imagine that each extends beyond the portion shown. Strip 6.3a was created by simply translating a basic design unit. The design unit is the repeating motif. The same design unit was used but it was reflected horizontally, thus creating a different motif. The symmetry group for the strip in b, therefore, contains horizontal reflec-

Figure 6.3. Each of the strips has a different symmetry group

tion as well as translation. It does not contain vertical reflection or rotation, as neither of these motions leaves the motif unchanged. In addition, however, it contains a glide reflect, as that would leave the strip's appearance unmodified. Looking next at Strip 6.3c the repeating motif again has two parts. This time, a glide reflect was used to obtain the second part. The strip's symmetry group, therefore, contains a glide reflect and a translation. Patterns a, b, and c then are characterized by different sets of isometries that leave them unchanged; that is, they have different symmetry groups. The motif in strip d was created by vertically reflecting the basic design unit, and e by rotating the unit 180°. (Here the lower right tip of the unit was used as the point of rotation,

159

but that is arbitrary. Different points of rotation would lead to different motifs but they would all have the same symmetry group.) For each of strips f and g, two motions were used to create the repeating motif. In f, the design unit was reflected both horizontally and vertically. This strip, however, has quite a number of symmetries: translation, vertical reflection, horizontal reflection, rotation, glide reflect, and the numerous combinations of these. Finally, g was created by a 180° rotation of the design unit followed by a vertical reflection of the result. Were a different design unit selected, the seven resulting strip patterns would look quite different. The point is, however, that regardless of the design, every repeating strip pattern falls into one of the seven categories depending on the logic of its construction and hence its symmetries.

To refer to the different symmetry groups, each needs a name or distinguishing identification. Because they were studied by many Westerners independent of one another, several different naming schemes exist. The scheme we use was developed by Russian crystallographers and is now accepted as the international standard. In this naming scheme, each of the seven groups is identified by a four-character symbol. If the

Table 6.1. Names of the symmetry groups for Figure 6.3

Strips in Figure 6.3	Construction used	Name	Another example
a	Translation	*p*111	PPPPPP
b	Horizontal reflection	*p*1*m*1	ᗞᗞᗞᗞᗞᗞ
c	Glide reflect	*p*1*a*1	PᑲPᑲPᑲ
d	Vertical reflection	*pm*11	PꟼPꟼPꟼ
e	Rotation 180°	*p*112	PᑯPᑯPᑯ
f	Horizontal/vertical reflection	*pmm*2	ᗞᑫᗞᑫᗞᑫ
g	Rotation/vertical reflection	*pma*2	PᑯᑲꟼPᑯ

160

group contains a translation, as every one of them does, the first character is the letter p. The second character depends on vertical reflection: m if it is a symmetry of the pattern or 1 if it is not. The next character refers to the horizontal: m if horizontal reflection is a symmetry or, if it is not, then a if glide reflect is a symmetry or 1 if it is not. Finally, the fourth place is 2 when a 180° rotation is in the group and 1 when it is not. The names for the strips in Figure 6.3 and another set of strips are summarized in Table 6.1.

So far we have confined our attention to design and have not involved color. Before we broaden to include color, let us look at a few Inca strip patterns in Figure 6.4. All of them, of course, include color, but the color defines or reinforces the design and so needs no special attention. We will discuss the Inca strip patterns later, but for now they are presented solely as illustrations: illustrations of strip patterns and illustrations of Inca creations.

When color is used in strip patterns, it can, as in the strips in Figure 6.4, be directly coordinated with the design. That is, if one part of a motif is, say, red in one repetition, then it is red in every repetition; and where it is, say, black, it is always black. The placement of these two colors underscores whatever symmetry is present. On the other hand, another coordination of color with design symmetries is, for example, if the colors are interchanged from one repetition to the next (that is, each time there is a translation) so that the motifs alternate in appearance rather than being exactly the same. Just two colors can be varied in numerous ways and the variations need not be coordinated with the design symmetries at all. In Figure 6.5a, for example, there is a coordinated color change: the color changes with each vertical reflection. In Figure 6.5b, the coloring and design symmetries are uncoordinated: the color alternates but does not do so in keeping with a design symmetry. When in coordination, the mathematical term is *perfect coloring*. Whether perfect or imperfect, the pattern, of course, is no better or worse. Perfect patterning, however, has a formal structure that is an extension of the design symmetry groups. And of particular importance is the insight that coloring can provide into the vision of symmetry held by the people who created the strips.

Because the color in Figure 6.5a changes with each reflection, we could say with some assurance that what we see as vertical symmetry of the motif is not just a concept we are introducing with our Western analytic approach. Figure 6.5b has vertical symmetry of the motif but, because of the coloring, we would have to look more carefully at what is happening. There is a single motif common to Figures 6.6a, b, and c; its symmetry group contains both horizontal reflection and a glide reflect, as well as translation. Each of the three different colorings, however,

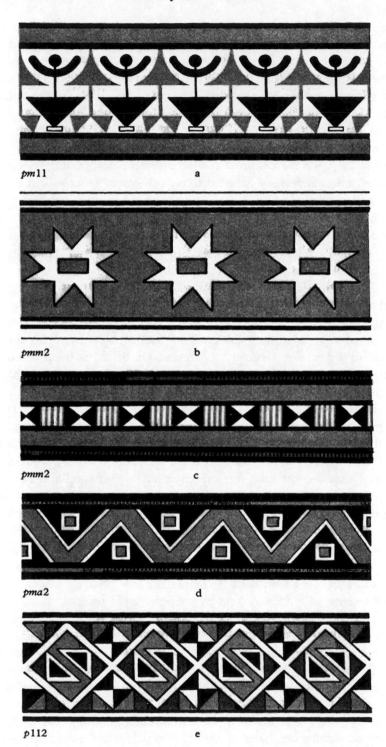

Figure 6.4. Inca strip patterns

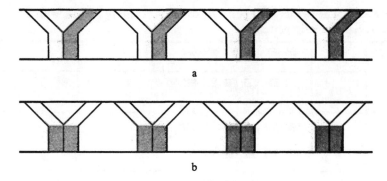

Figure 6.5. Color changes: a, coordinated with design symmetry; b, uncoordinated with design symmetry

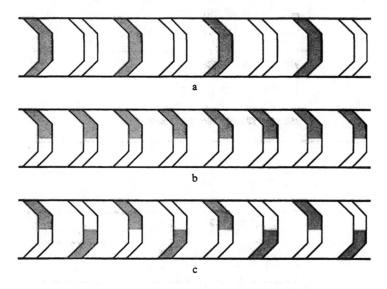

Figure 6.6. Three different perfect colorings

makes just one of those three symmetries prominent; each suggests a different emphasis.

As in the last example, several different perfect colorings can be associated with each of the seven design symmetry groups. Confining ourselves to just two variable colors, there are, in all, twenty-four perfectly colored symmetry groups: the seven already discussed, which have point-by-point replication of color, and seventeen additional ones involving color reversals. Their naming follows the same scheme as was used before. Now, however, a prime is added to a symbol when a change in color goes along with the motion it represents. The design symmetry

163

Table 6.2. The twenty-four perfectly colored symmetry groups, with an illustrative set of strip patterns

p111		⌐⌐⌐⌐⌐⌐
	p'111	⌐⌐⌐⌐⌐⌐
$p1m1$		⊏⊏⊏⊏⊏⊏
	$p'1m1$	⊏⊏⊏⊏⊏⊏
	$p1m'1$	⊑⊑⊑⊑⊑⊑
	$p'1a1$	⊏⊏⊏⊏⊏⊏
$pm11$		⌐⌐ ⌐⌐ ⌐⌐ ⌐⌐
	$p'm11$	⌐⌐ ⌐⌐ ⌐⌐ ⌐⌐
	$pm'11$	⌐⌐ ⌐⌐ ⌐⌐ ⌐⌐
$p1a1$		⌐⌐ ⌐⌐ ⌐⌐ ⌐⌐
	$p1a'1$	⌐⌐ ⌐⌐ ⌐⌐ ⌐⌐
$p112$		⌐⌐ ⌐⌐ ⌐⌐ ⌐⌐
	$p'112$	⌐⌐ ⌐⌐ ⌐⌐ ⌐⌐
	$p112'$	⌐⌐ ⌐⌐ ⌐⌐ ⌐⌐

Table 6.2 (continued)

pmm2		
	p′mm2	
	pm′m2′	
	pmm′2′	
	pm′m′2	
	p′ma2	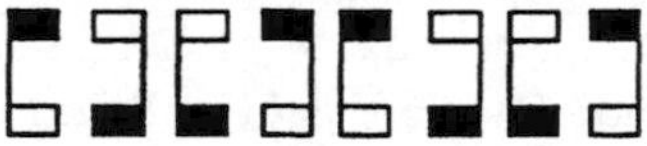
pma2		
	pma′2′	
	pm′a′2	
	pm′a2′	

group of Figure 6.5a is $\underline{pm11}$; its colored symmetry group is $\underline{pm'11}$ since the color changes with each vertical reflection. For Figure 6.6, the design symmetry group is $\underline{p1m1}$ and so the colored symmetry groups are (a) $\underline{p'1m1}$, (b) $\underline{p1m'1}$, and (c) $\underline{p'1a1}$. (This last symbol required another modification as well because the glide reflect was subsumed by the horizontal reflection in the original $p1m1$). Table 6.2 lists the twenty-four perfectly colored symmetry groups with an illustrative set of strip patterns. We will refer back to these ideas as we look at some configurations created by the Maori and by the Inca.

165

The construction of a *marae*, a meeting house that serves as the center of community activity, calls upon and unites the traditional Maori crafts and decorative arts. The walls alternate panels of wood carvings and latticework panels with cross-stitched designs. Carvings adorn the sills and lintels of the doors and windows and the barge-board at the roof line. Flax mats are made for the floor, special cloaks are woven for the major guests who will come, and bodices and skirts are woven for the people who will welcome the visitors. The rafters and ridgepole of the *marae* are decorated with a particular type of design in particular colors. The designs, *kowhaiwhai*, are painted on the wood or are carved in low-relief and then painted. Two master craftspeople, usually a man and a woman, are chosen to oversee the project. The man designs and supervises the carvings and *kowhaiwhai* and the woman designs and supervises the latticework and weaving. Except for the execution of some major pieces, the community constructing the *marae* does most of the work under the instruction and close scrutiny of the master craftspeople. Master craftspeople are few and, as a result, most often are not from the immediate community. They come from another place and sometimes must stay as long as three or four years until the *marae* is completed. They are paid well and given, in addition, many gifts and considerable honor and respect.

The *kowhaiwhai*, often referred to as rafter patterns, are the strip patterns that interest us, and it is the master carpenter who is their creator and who carries them on as part of the carving tradition. By the 1920s, the number of master carvers had dwindled. Schools were established, across regional lines, to ensure that the tradition would continue. Whether in the schools or before their advent, the learning and transmission of the craft occur primarily through apprenticeship and at the time a *marae* is being constructed. Designs, styles, techniques, knowledge about the quality and preparation of wood, lore and history, and the rules and behaviors associated with carving all are part of what is taught. Attitudes and observances are a necessary part of the technical skill because the community center, and carvings in general, must be ritually correct. It is, for example, forbidden to use chips or shavings from the carvings for fuel or to remove shavings by blowing on them. Hands must be washed before and after work. And the carvings and tools must not be contaminated by the presence of women, fire, or cooked food.

The rafters of a *marae* are flat on their top side and curved on the bottom. The *kowhaiwhai* is first outlined with a chisel-like tool and then painted. While planning the design, the carver does not use any intermediate model or medium; that is, he does not sketch all or part of the pattern on, say, a piece of paper and then copy from that onto a rafter. The carver, therefore, must visualize the overall pattern and the appropriate shapes and sizes of the various parts that will enable his visualization to be transferred onto a *specific* piece of wood. It is a particularly crucial ability when creating rafter patterns with symmetries that depend on precise interrelationships. A contemporary manual on traditional Maori carving includes instructions for a shape referred to as a scroll or *pitau*. The scroll is described as one of the most common components of surface decorations and, in particular, as the most common element of rafter designs. The shape (shown in Figure 6.7) has a stem and a bulb, and the spaces around it are often filled with crescent-like texturing. Examination of rafter patterns indeed show that many of the motifs have been created by combinations and juxtapositions of this one element. Examples can be seen in Figure 6.8, a photograph taken prior to 1897, of the porch area of the Ngati-Porou *marae*. The walls of carved panels and decorated latticework are visible, as well as three or four decorated rafters. As with all rafter patterns, the colors are white, black, and red; the curvilinear design is in the white of the wood and its background, varying between black and red, is usually pigmented by soot and red ochre. The symmetries of the *kowhaiwhai* are striking. One traveler observed that in a single case where he saw no symmetries on a set of rafters, he finally realized that symmetry was nonetheless present when each rafter was viewed in relationship to the one opposite with respect to the center ridgepole.

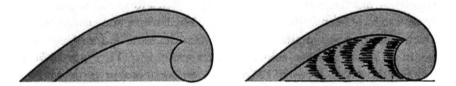

Figure 6.7.   The scroll and texturing around the bulb

Let us look in detail at some *kowhaiwhai* patterns by utilizing our Western symmetry analysis. First, in Figure 6.9a, look at the repeating design motif while disregarding the colored background. Were it not for the individual scrolls just adjacent to the vertical lines, the motif would have vertical, horizontal, and rotational symmetry so that the symmetry

Figure 6.8.   The Ngati-Porou *marae*

group of the strip would be *pmm*2. Those side scrolls, however, are rotated forms of each other and so the horizontal and vertical symmetries appear to have been deliberately broken. The background color reverses from repetition to repetition, that is, when the motif is translated, and so, in all, the strip's two-color symmetry group is *p*′112. Figure 6.9b is again in the design group *p*112 (rotation, no horizontal or vertical reflection) but this time the color reiterates the design; both red and black rotate or translate into themselves in keeping with design rotation or translation. A third rafter pattern with the design group *p*112 is shown in Figure 6.9c. It is another example in which small scrolls change the symmetry of the larger design. Were it not for the small scrolls, the design symmetry group would be *pma*2, including vertical reflection and glide reflect, as well as rotation. However, here, too, the scrolls are rotated versions of each other and so break the symmetries by reflection. This time, the colors reverse with each rotation and so the two-color symmetry group is $p112'$. Thus, in these three rafter patterns, we see each of the three possible colorings of a strip with design symmetry group $p112$, but we also see small designs with restricted symmetry embedded in larger designs with a greater number of symmetries.

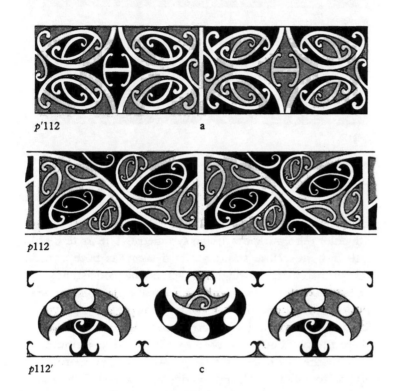

Figure 6.9. Maori rafter patterns

Figures 6.10a and 6.10b are interesting to compare; they have the same basic design units combined differently and colored differently. In Figure 6.10a, the design unit is rotated and vertically reflected and the color is preserved for each design symmetry (two-color symmetry group *pma2*). In Figure 6.10b, on the other hand, the design unit is just rotated; the color is preserved on the rotation but reversed for each appearance of the motif (two-color symmetry group $p'112$).

A few more examples of the rafter patterns are included to give a greater sense of their style and variety. In Figure 6.11a, the design group is $p1a1$ and color reversal takes place with each glide reflect (two-color

pma2 a

p'112 b

Figure 6.10. Two Maori rafter patterns with the same basic design unit

symmetry groups $p1a'1$). Figure 6.11b has a design symmetry group of $p1m1$ with color reiterating the design symmetry. The pattern in Figure 6.11c again requires close scrutiny. Its design has both vertical and rotational symmetry. Its coloring, however, neither consistently stays the same nor consistently reverses with the vertical reflection. It does, however, consistently reverse upon rotation (group $p112'$). In two earlier examples, apparent vertical symmetries were broken because of small scrolls; here the symmetry is broken by the overall coloring. Our final example is another with an interesting anomaly. The motif in Figure 6.11d has horizontal, vertical, and rotational symmetry. The coloring, at first glance, repeats these symmetries and reverses red and black from repetition to repetition. However, if the color alternation of the elliptical figure is compared with the color alternation of the small triangles near the edges, they can be seen to be offset from each other. Or, put another way, the horizontal and vertical reflections of the elliptical part preserve the color, while the triangle's colors alternate upon horizontal and vertical reflections. In our Western categorization, the design symmetry of the strip is reduced (from group $pmm2$ to $p'112$), but, alternatively, we could say that the underlying design symmetry remains and the coloring increases the interest of the strip by juxtaposing one symmetry with another.

Clearly the *kowhaiwhai* are formally structured. A design on a strip need not have any symmetry at all. Some few of the motifs are said to evoke certain aspects of the natural environment for the Maori; one, for example, suggests natural food resources, and another the ocean. Of

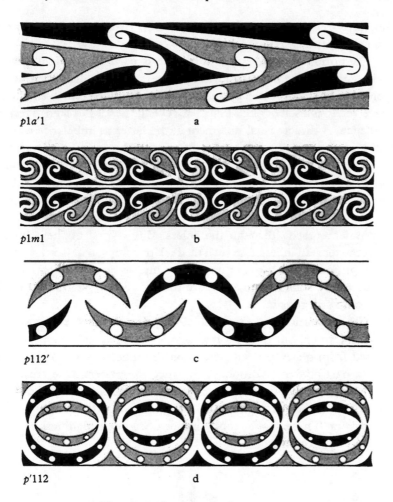

pla'l   a

plm1   b

p112'   c

p'112   d

Figure 6.11. Maori rafter patterns

deeper significance, however, is their underlying structures with their harmonies, balances, rhythms, symmetries, and asymmetries. One cogent theory views these structures as related to, and expressive of, the structures that underlie the Maori belief system. Complementarity, the relatedness of pairs through difference, and symmetry, the relatedness of pairs through sameness, are seen as organizing principles in much of Maori myth, religion, social life, and economics. The male-female complementary relationship is fundamental, as is the complementary relationship between the world of humans and the world of the gods. All events are seen as having both physical and spiritual explanations and earthly as well as cosmic significance. These complementary relationships have a dynamic component: union is followed by separation is followed by reunion, and so on. Many Maori rituals serve to maintain the stability of these relationships.

The notion of symmetry and balance is particularly significant in social relations. One expression of it is the exchange of food which is supposed to be reciprocated by an equivalent. Similarly, when an offense is perceived to oneself or one's kinsmen, revenge needs be taken to restore the balance. A direct visual statement of the latter as imbalance was the practice of shaving one side of the head until the revenge was accomplished. In this interpretation, the underlying Maori structure of reality is one based on symmetric pairs and complementary pairs, which are both in tension as balances are broken and restored. Seen from this point of view, the rafter patterns express the symmetry through the designs and the complementarity through the colors. But several of them further express the tension inherent in such dual relationships; that is, an apparent overall symmetry is broken or made asymmetric by a small part within it. Even where such asymmetries are present, other symmetries in the relationship still remain.

The conception and execution of the rafter designs are the work of special craftsmen, but their individual work is a part of a tradition that is passed from generation to generation. Symmetric strip patterns are clearly a part of that tradition. Also, since the patterns are created to adorn a building that is the pride of a community and the focal point of its activities, interest in symmetric strip patterns extends well beyond their makers. These spatial configurations become a significant part of the Maori environment.

4 For the Incas, we will look at strip patterns painted on pottery. As compared to the Maori rafters, the objects are different, the material is different, the style of the motifs are different, and, of course, the culture is different. As with the Inca quipus (Chapter 1), the pottery pieces are archaeological artifacts. They come from widely scattered graves and were probably used as we use dishes, cups, bowls, and containers. Because of the relatively early demise of Inca culture, we have little useful information about the craftspeople who made the pottery designs. We can, nonetheless, observe that the strip patterns are marked by formality, precision, and repetition, with some symmetries more prevalent than others. In addition, we can relate aspects of these patterns to other archaeological evidence and, even more important, see how these strips both reflect and express characteristics of Inca culture.

The walls of many state buildings constructed by the Incas still stand. The patterns created by the assemblage of stones forming the walls give a general impression of linear outlining, formality, and regularity. The precise way that immense stones are fitted together is, perhaps, their most impressive feature. The Incas were familiar with sun-dried bricks and bonding materials as they used them in the construction of ordinary buildings. But, for state buildings, they chose instead to use large, heavy, shaped stones. The shapes of the stones in a wall are not uniform. They all, however, are angular geometric forms set so tightly together and aligned so exactly that nothing is needed between them to fill gaps or to serve as an adhesive. Each stone is beveled back toward its edges so that its face appears as an angular geometric figure set forward of the same figure with larger dimensions (see Figure 6.12).

Another, quite different example of repetition, regularity, and precision is the layout and construction of a typical agricultural settlement called Inty Pata. The settlement was on a very steep mountainside; the slope of the portion with the settlement exceeds 40°. Four compounds made up the residential area. In Figure 6.13, you can see that the compounds were arranged in two repeated rows of two compounds each. Within each row, the compounds are laid out as vertical reflections of each other. (Each compound has a corner storeroom, two units facing each other across a courtyard, and another open unit facing the courtyard.) In order to farm such a piece of land, more than fifty terraces were constructed, each made up of a stone wall behind which coarse sand and clay were deposited and then topped with fine surface soil. Stairways were built from the residential areas to the terraces and from terrace to terrace; water channels were built to conduct the water carefully down the mountain to each terrace. What we see is not only evidence of very difficult work but also evidence of very detailed and precise spatial planning. The layout conforms to the contour of the mountainside, but, at the same time, superimposes a rigorous structure upon it. We said before, in the context of the quipus, that the Incas were methodical and concerned with detail; they also had concern for formally structured spatial arrangements.

Our comments regarding Inca pottery decorations are based on examination of approximately 250 strip patterns. To begin with, all of these are repeated motifs along a strip and, in addition, have only one or two variable colors. For the Maori patterns, there were symmetries within symmetries and close scrutiny was needed to fully savor the internal relationships. The Inca strips are more straightforward. By and large (that is, by over 90 percent), they are perfectly colored, with the vast majority having colorings that reiterate the design symmetries rather

Figure 6.12. A girl at an Inca wall, Cuzco, Peru, 1975

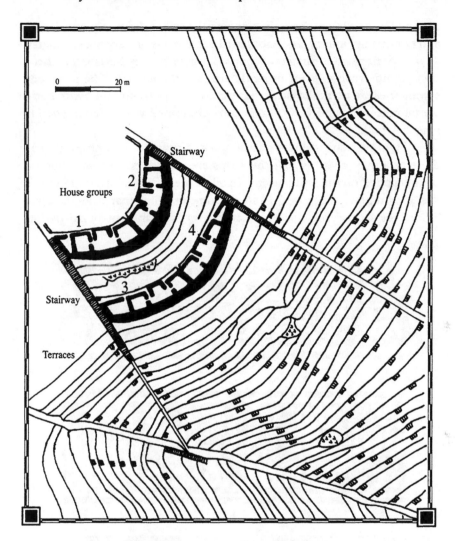

Figure 6.13. The layout of Inty Pata

than adding variety to them. Just about 10 percent show only translation and no other symmetry, and another 10 percent have rotational symmetry as well. Strips with symmetry groups characterized by horizontal reflection or glide reflect (*p1m1* or *p1a1*) are remarkably rare. The prevalent symmetries are vertical reflection, vertical reflection combined with horizontal reflection, and vertical reflection combined with rotation (*pm11*, *pmm2*, *pma2*).

Figure 6.4 showed several examples. A characteristic property that these examples share is that the strips are clearly defined as strips by being banded on top and bottom. In all of these, the color reiterates the design symmetries. Figure 6.14 shows some in which that is not the case. The three strips each have horizontal, vertical, and rotational design

pm'm'2   a

p'mm2   b

p'ma2   c

Figure 6.14. Inca strip patterns

symmetries (*pmm*2). For all the designs, this symmetry group is the most prevalent, accounting for over one-third of them. It is also the category for which there are the most varied colorings. With their colorings considered, Figures 6.14a and 6.14b are in groups *pm'm'*2 and *p'mm*2, respectively. For Figure 6.14c, the overall strip pattern is formed from two bands. They each have the same motif and each alternates color from motif to motif. The colors of the bands, however, are offset so that as a unit there is alternation from top to bottom as well. (In terms of their two-color symmetry groups, each band is *p'mm*2; but when combined, the complete pattern is *p'ma*2.) The creation of a pattern by combining two similarly designed bands is not uncommon on these pottery pieces and we will look at a few more of them. They are strongly reminiscent of the residential layout at Inty Pata, both because of the vertical symmetry within each row of houses and because of the repetition of the row (look at Figure 6.13 again).

Additional Inca strip patterns are shown in Figure 6.15. These have the interesting feature of utilizing the same basic geometric forms— isosceles trapezoids and isosceles triangles—with different juxtapositions and different colorings. There is a single band and a double band, each with no color alternation; and then four double bands, each with a different coloring. And, the double bands are formed in two different ways from the single bands. Let us look more closely at the six strips. Figure 6.15a has rotational symmetry, as well as vertical symmetry and no color variation (*pma*2). Figure 6.15b is formed from two bands, one above the other, but relatively positioned so that the combination has a horizontal line of symmetry. The vertical symmetry still remains, as does the rotational symmetry, although now with a different point of rotation. The patterns on strips c and d have the same construction as b, whereas e and f do not. In e and f, the pair of bands repeat rather than reflect each other. Summarizing the six patterns, ignoring their coloring for the moment, the single band pattern is of type *pma*2, but when two single bands are combined into a double band pattern, the result is either *pmm*2 as in b–d or remains *pma*2 as in e and f. Next we include the color. In each of the bands of b, the colors reiterate the design symmetries and each band has the same coloring. In their combination, then, the colors reiterate the design symmetries as well. In c, on the other hand, the colors of the two bands interchange with the horizontal reflection (*pmm'*2'). Both bands in d, when viewed separately, have identical colorings. Individually or as a unit, the pattern to their coloring scheme is that there is alternation from one repetition of the motif to the next (*p'*111 for each, *p'ma*2 for the pair). The coloring in e is similar, but not quite the same.

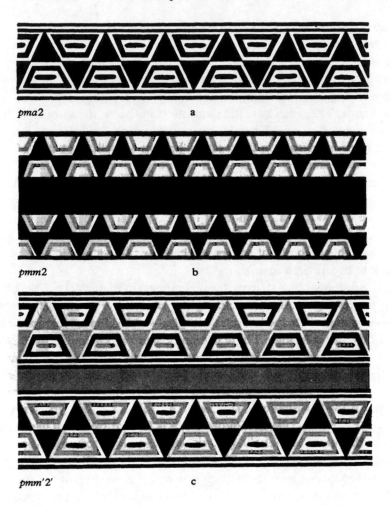

*pma*2   a

*pmm*2   b

Figure 6.15. Inca strip patterns

*pmm'*2'   c

The design and coloring of the two bands making up the strip are identical, but the design symmetries of rotation and vertical reflection are not borne out by either consistent repetition or consistent reversal of the colors. However, just as for the strip before, color reversal takes place from motif to motif (*p'*111). The last strip in the set is again a quite different coloring of the bands; this time, there is consistent interaction between design symmetries and color—for the individual bands or paired bands, the colors stay the same upon reflection and translation and reverse upon rotation and glide reflect (*pma'*2').

It is tempting to see in this set of strips an exploration of possibilities; that is, the intentional creation of single bands of type *pma*2 combined variously into double bands of *pmm*2 or *pma*2 and these colored

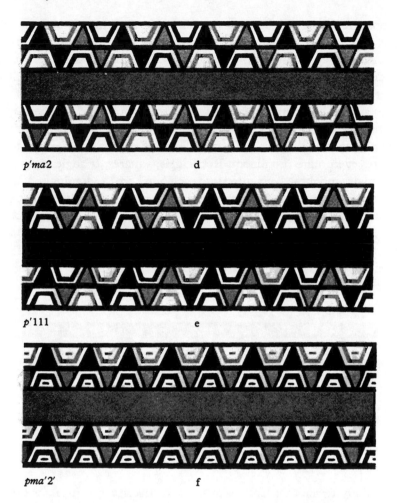

p'ma2   d

p'111   e

pma'2'   f

variously to become *pmm2*, *pmm'2'*, *p'ma2*, *p'111*, or *pma'2'*. While this is most probably not the case, we can at least say that the simple trapezoidal form *was* elaborated into a variety of formal, repetitious, and symmetric strip patterns. The widespread use of a single form or its reuse with minor variations is distinctive of the Incas. This was one of the ways they impressed and maintained the fact of their presence on the diverse cultures they controlled.

In all, for this collection of strip patterns, double reflection is most prevalent and next are those with rotational and vertical symmetry or vertical symmetry alone. The coloring, in general, is consistent with the design symmetry, mostly by remaining the same, but sometimes by alternation. When two colors are varied, the seven design symmetry groups can be augmented by seventeen more. For these Inca strip patterns, only nine of those seventeen categories were present, but, more important, they involved only about one-eighth of the strips. As such, the design is paramount and the use of color is conservative, doing little to introduce visual variety. The sense of formality and regularity of the strip patterns is not only from their symmetries but also from the shapes and figures that constitute them. The shapes are rectilinear and, similar to what is seen on the Inca walls, outlining makes their linear and angular features even more prominent. Rectangles, squares, triangles, isosceles trapezoids, and these segmented into each other are the dominant figures. As a last example of an Inca strip pattern, Figure 6.16 epitomizes these characteristics and provides extreme contrast to the sweeping, curvilinear forms of the Maori. In Figure 6.16, a spiral is linearized, made rectangular, and combined with triangles embedded in squares. Each spiral pair is distinctly enclosed and separated from the others and, again, there is a clear border delineating the fact that it is a strip pattern.

Figure 6.16. An Inca strip pattern

Many terms that have been used to describe the Inca are exemplified by their strip patterns. The Incas have been described as methodical and conservative, with concern for detail, concern for precision of fit, and concern for spatial arrangements. Here, we can see a specific expression of those characteristics in a particular type of configuration. We see formal spatial ordering not only in the creation of symmetric strips but also in the use and interconnection of fundamental geometric shapes.

5 It is quite easy to distinguish the strip patterns of the Maori from those of the Incas. Figures 6.4e and 6.9b, for example, both have the same symmetry group, but are considerably different in style, as each is a part of the culture from which it comes.

They both, however, evidence a pervasive human concern, namely, the ordering of space.

There is, as we said before, no need to make any orderly decorations and certainly no need to make decorations with such a high degree of internal organization. But people do. And, what is more, the patterned nature of decorations is not isolated from other aspects of a culture. Whether they are curvilinear or rectilinear, these strips evidence mental and physical manipulation of shapes and forms within the self-imposed constraint that they must repeat along a strip. They do not enhance the practical function of what they adorn, and, as such, the creation of these spatial patterns and their appreciation by others in the culture who do not create them demonstrates an interest in spatial ordering.

Notes

2. The study of symmetric patterns in one, two, and three dimensions has attracted a variety of scholars. In the nineteenth century, it was primarily crystallographers who were interested in ascertaining all possible forms of repeated patterns in three dimensions. In 1891, a list of the 230 possibilities was published by E. S. Federov. In the same year, in another paper, he listed the seventeen possible two-dimensional patterns (*Proceedings of the Imperial Saint Petersburg Society*, series 2, volume 28, pp. 1–146 and pp. 345–389, respectively). A textile physicist, H. J. Woods, in "The geometrical basis of pattern design. Part I," *Transactions, Journal of the Textile Institute*, 26 (1935) 197–210, first discussed two-color symmetry for one dimensional patterns (strip patterns). In the 1920s, the mathematicians G. Pólya and A. Speiser united the crystallographers' study of pattern with the mathematical concept of symmetry groups. (G. Pólya, "Uber die Analogie der Kristallsymmetrie in de Ebene," *Zeitschrift für Kristallographie*, 60 (1924) 278–282 and A. Speiser, *Theorie der Gruppen von endlicher Ordnung*, 2nd edition, Springer, Berlin, 1927). A student of Speiser's, E. Müller, applied the ideas to two-dimensional decorations from another culture, *Gruppentheoretische und Structuranalytische Untersuchungen der Maurischen Ornamente aus der Alahambra in Granada*, Ph.D. thesis, University of Zürich, Rüschlikon, 1944. My own introduction to the subject was the excellent work of the archaeologist A. O. Shepard, *The Symmetry of Abstract Design with Special Reference to Ceramic Decoration*, Contributions to American Anthropology and History, vol. 9, no. 47, Carnegie Institution of Washington, Publication 574, Washington, D.C., 1948, which was stimulated by G. W. Brainerd's "Symmetry in primitive conventional design," *American Antiquity*, 8 (1942) 164–166. They introduced the mathematical analysis of symmetry into the study and classification of archaeological artifacts.

Perhaps because of the classic *Symmetry*, H. Weyl, Princeton University Press, Princeton, 1952 (first delivered as a series of four lectures at Princeton in 1951), since 1960 mathematicians have become increasingly interested in symmetry. There have been extensions in theory and extensive use of examples from other cultures. Also, there have been several excellent summaries and expositions of the results found in diverse sources. Specifically about the generation of one- and two-color strip patterns, D. Schattschneider's, "In black and white: How to create perfectly colored symmetric patterns," *Computers and Mathematics with Applications*, 12B (1986) 673–695, is recommended. In a more theoretical vein, "Colored frieze groups," J. D. Jarratt and R. L. E. Schwarzenberger, *Utilitas Mathematica*, 19 (1981) 295–303, deals with N colors. They prove the interesting result that the number of different strips is 7 for N odd, 17 when $N = 2$ (mod 4), and 19 when $N = 0$ (mod 4). A simple proof of their result is in H. S. M. Coxeter's "A simpler introduction to colored symmetry," *International Journal of Quantum Chemistry*, 31 (1987) 455–461. The book *Symmetries of Culture*, D. K. Washburn and D. W. Crowe, University of Washington Press, Seattle, 1988, provides an excellent bibliography, many illustrations, and clear mathematical exposition. Its focus is symmetric patterns in one or two dimensions and in one or two colors. The book *Tilings and Patterns*, B. Grunbaum and C. G. Shephard, W. H. Freeman and Co., New York, 1987 deals more generally with the description and analysis of pattern.

3. *The Art Workmanship of the Maori Race in New Zealand*, Part II, A. Hamilton, New Zealand Institute, Wellington, 1897, contains twenty-nine rafter patterns as well as a description of the building of a central community house. (Hamilton attributes these to the Rev. Herbert Williams.) I acknowledge with appreciation the permission of The Royal Society of New Zealand to reproduce some rafter patterns. In 1977, the book was reprinted by New Holland Publishers, London, entitled *Maori Art*. I acknowledge their permission to reproduce the photograph in Figure 6.8. Additional information on the building of a *marae* and the tradition of the Maori carvers is in "The Maori carver," R. W. Firth, *Journal of the Polynesian Society*, 34 (1925) 277–291; R. Linton, "Primitive art," *Kenyon Review*, 3 (1941) 35–51; S. M. Mead, *The Art of Maori Carving*, A. H. & A. W. Reed, Wellington, N.Z., 1961; and pp. 273–282, in J. Metge, *The Maori of New Zealand Rautahi*, Routledge & Kegan Paul, London, revised ed., 1976.

Several people have applied Western symmetry analysis to the rafter patterns published by Hamilton. G. Knight in "The geometry of Maori art-rafter patterns," *New Zealand Mathematics Magazine*, 21 (1984) 36–40, shows one from each of the seven design symmetry groups. His examples are Hamilton's nos. 8, 10, 16, 26, 28, 29, and one beam from the photograph in Figure 6.8. J. D. H. Donnay and G. Donnay in "Symmetry and antisymmetry in Maori rafter designs," *Empirical Studies of the Arts*, 3 (1985) 23–45, analyze all twenty-nine of Hamilton's patterns in terms of two-color symmetries. Washburn and Crowe in *Symmetries of Culture* (cited above) include two-color analysis of Hamilton's nos. 6, 10–15, 19, 22–23, 25, and 28, as well as of two additional patterns (their Figures 4.6 and 4.9) from H. Bossert's *Folk Art of Primitive Peoples*, Praeger, N.Y., 1955. The examples used here are from Hamilton. My Figures 6.9, 6.10, and 6.11 are, respectively: nos. 22, 6, 12; 20, 19; and 23, 28, 13, 11.

The writings of F. A. Hanson place the mathematical analyses of the rafter patterns in the broader context of Maori art and culture. It is F. A. Hanson who observed the paired rafters in a house which now is in the National Museum in Wellington. Photographs of three rafters from this house are in his "From symmetry to anthropophagy: The cultural context of Maori art," *Empirical Studies of the Arts*, 3 (1985) 47–57, and four more are in his "Art and the Maori construction of reality," in *Art and Artists of Oceania*, S. M. Mead & B. Kernot, eds., Ethnographic Arts Publications, Mill Valley, California, 1983, pp. 210–225. These two articles and his "When the map is the territory: Art in Maori culture," in *Structure and Cognition in Art*, D. Washburn, ed., Cambridge University Press, N.Y., 1983, pp. 74–89, and his (with L. Hanson) *Counterpoint in Maori Culture*, Routledge & Kegan Paul, Boston, 1983, present and support in detail the cultural structural analysis cited here. "Art and the Maori construction of reality" is particularly recommended.

4. The Inca strip patterns that were analyzed are from *Motivos de Ornamentacion de la Cermica Inca-Cuzco*, vol. 1, J. F. Baca, Libreria Studium S.A., Peru, 1971, an authoritative collection which its author worked on for over forty years. I am indebted to Carole Frick for her assistance with the drawings in Figures 6.4, 6.14, 6.15, and 6.16. They are based, respectively, on the following figures in Baca's book: 339, 717, 170, 249, and 499; 371, 184, and 582; 267, 266, 268, 715, 270, and 269; and 520. In addition to this analysis of one- and two-color strip patterns, 120 symmetric planar decorations taken from the same book were also analyzed. The mathematics of symmetric planar decorations, frequently referred to as *wallpaper patterns*, has not been included in this book. There are, however, seventeen one-color plane symmetry groups. (A discussion of plane symmetry is included in the book by Washburn and Crowe cited above and the article "The plane symmetry groups: Their recognition and notation," D. Schattschneider, *American Mathematical Monthly*, 84 (1978) 439–50, is also recommended.) That analysis also showed the Incas to be conservative in their design vocabulary when extending from one-dimensional strips to the two-dimensional plane. Of the seventeen possible plane symmetry groups, five were never present, while four accounted for 70 percent of the surface decorations.

The comments about the Inca are based on the much more detailed discussion in *Code of the Quipu*, M. Ascher and R. Ascher, University of Michigan Press, Ann Arbor, 1981. The pattern analysis included here is an extension to two colors of the analysis discussed there (pp. 54–55). Figures 6.12 and 6.13 appear in that book as well. Figure 6.13 was adapted by Robert Ascher from *Archaeological Explorations in the Cordillera Vilcambamba Southeastern Peru*, P. Fejois, Viking Fund Publications in Anthropology, No. 3, Wenner-Gren Foundation for Anthropological Research, New York, 1944.

In Conclusion: Ethnomathematics

s   e   v   e   n

Mathematical ideas involve number, logic, or spatial configuration and, in particular, the combination or organization of these into systems or structures. In the previous chapters, we ranged broadly both in terms of these ideas and in cultural settings. We have, however, only begun to explore the worldwide instances of the diverse expressions of mathematical ideas. But, with these examples in hand, we can reflect more generally on them and on ethnomathematics.

As a whole, mathematical ideas are rich and multifaceted. There is no particular single path along which they must develop in every culture; they cannot be ordered or compared along any single linear scale. A culture's use of numeral classifiers, for example, does not come before or after another culture's lack of them; the Navajo concept of spacetime is neither better nor worse than that of Western culture; and the Warlpiri organization of kinship involving a dihedral group of order 8 is neither ahead of, nor behind, our own. A single straight line is far too simple to serve as an image of how all these mathematical ideas are related to each other. For me, a possible visualization of their totality is the revolving, lighted, mirror-faceted globe that is suspended from the ceiling of some ballrooms. Each of the thousand or so small mirrored facets is contiguous to some, but widely separated from others. Which facet catches and reflects the light at any given moment depends on where you are in the room.

In all of the examples, as they have been presented, there is an interplay of mathematical ideas and culture. They cannot be separated from each other. That a culture uses numeral classifiers and what the classifications are tells us simultaneously of the people's ideas regarding number, something of their language, and something about how they categorize the world around them. Similarly, the sand tracing tradition of the Malekula simultaneously involves myths related to death, an interest in tracing continuous figures, specific figures that were created, and logical procedures for the tracing of the figures. In Western culture and among the Tshokwe of Africa, the cultural surroundings of the graph theoretical ideas are not the same, nor should we expect that they would be. The strip patterns that we see around us, those of the Incas, and those of the Maori are quite different in style, in usage, and in their other cultural linkages. Shared is the creation of strip patterns and an interest in them, but not necessarily shared is the motivation for their creation, nor the world view or aesthetic that leads to the particular strips that result. Our Western mode of pattern analysis and classification was used to suggest some insights into the patterns of others that perhaps we would not have seen otherwise. What we term mathematical ideas are, in every instance, part of the web that constitutes a culture. Hence, as we study these cases, we draw upon, but also add to, the understanding of other peoples.

And, whenever we increase our understanding of other cultures, we increase understanding of our own by seeing what is or is not distinctive about us and by shedding more light on assumptions that we make which could, in fact, be otherwise. Our concepts of space and time are, after all, only our ideas and not objective truth. And, there is no single correct way to depict objects in space, nor one correct way to orient a picture in order to comprehend its contents.

Our purpose, however, is more specific than simply to increase understanding of diverse cultures. We are concerned with a particular aspect of human ideation, namely, mathematical ideas. Such ideas exist in all cultures, but which ones are emphasized, how they are expressed, and their particular contexts will vary from culture to culture. The Incas, for example, were intensive data users and so number, logic, and spatial configuration became uniquely combined in their quipus. Social relations are of paramount importance to the Warlpiri and so it is there that elaborate logical structuring is found. In contrast to both, the Malekula combined logical structured processes with the creation of intricate spatial forms. For each of these groups, we have touched upon just one or two of their mathematical ideas; they surely have others. In any case, no culture has every idea and, from culture to culture, the appearance of even the same idea will vary.

There is value in focusing on mathematical ideas across cultures because these are human mental constructs and, as such, they can add to our appreciation and understanding of what sets humankind apart from other animals or from inanimate objects. To us, mathematical ideas are an important aspect of what it means to be human.

Since it is not possible to clearly delineate the scope of mathematical ideas or to draw sharp, clear lines between these ideas and other ideas, this area of human thought clearly overlaps with other areas of thought. But that is certainly true of any divisions that are made to organize any areas of study. There are no hard and fast lines between literature and history, between chemistry and physics, between psychology and sociology, or between any other disciplines. And, in looking across cultures, there is no clear demarcation between studies of kinship and studies of religion, or between studies of art and studies of mythology. Each is an organizing device that highlights one dimension for in-depth viewing and each is meaningful as long as it is understood that some aspects of it can and should fall into other areas for other viewings. Because there are people in our culture who are particularly interested in mathematical ideas, and because many people place special value on these ideas and on those who think them, knowledge about their existence and expression across cultures becomes very important.

As we seek these ideas across cultures, we recognize that even within our own culture, mathematical ideas exist in many different contexts. Thus, they are not the exclusive province of some unusual few who were specially endowed at birth or specially educated. And, as Western education, including the Western conceptualization and expression of mathematical ideas, becomes spread throughout the world, it becomes increasingly important that the other expressions are recognized and valued and not lost to all of us.

4 Ethnomathematics, as it is being addressed here, has the goal of broadening the history of mathematics to one that has a multicultural, global perspective. It involves the study and presentation of mathematical ideas of traditional peoples. This broadening of perspective to include other cultures has the associated effect of enlarging the history of mathematics from dealing primarily with the Western professional class called mathematicians to involving all sorts of people. In some of our examples, different professional groups were involved, and in several others we discussed ideas of the cultures at large. For the Caroline Island spatial model, the ideas were limited to a professional group called navigators. For the Inca quipus as well, there was a select group whose special training included the representation and use of numbers. And the symmetric rafter patterns of the Maori were the work of special craftspeople. The number words of the different cultures and kin system of the Warlpiri, on the other hand, are known throughout those cultures. The probabilistic, logical, and strategic ideas within the native American and Maori games and African puzzles are in the province of all in the culture but probably of more interest and enjoyment to some than to others. And the spatial concepts we discussed were the ideas of the Navajo people and Inuit people, whereas we noted—and this is extremely significant—that in Western culture there is a difference between the space-time ideas of some professionals and those of the general populace.

There are, in every culture, groups or individuals who think more about some ideas than do others. For other cultures, we know about the ideas of some professional groups or some ideas of the culture at large. We know little, however, about the mathematical thoughts of individuals in those cultures who are specially inclined toward mathematical ideas. In Western culture, on the other hand, we focus on, and record much about, those special individuals while including little about everyone else. Realization of this difference should make us particularly wary of any comparisons across cultures. Even more important, it should encourage finding out more about the ideas of mathematically oriented innovators in other cultures and, simultaneously, encourage expanding the scope of Western history to recognize and include mathematical ideas held by different groups within our culture or by our culture as a whole.

A truly global and humanistic history must extend, not only to other cultures, but beyond a few select groups and individuals to ideas that pervade a culture.

5 In addition to broadening the history of mathematics, our goal also includes modification of some notions that are currently found within it. Although references to traditional peoples have been infrequent in the mathematical literature, those that do appear are most often underpinned by theories of culture that were dominant in the late nineteenth and early twentieth centuries but that no longer prevail. We would prefer to simply continue beyond them, but we explicitly discuss them here because they have skewed our view of traditional peoples by implying that their mathematical ideas could only be minimal or of little interest.

According to the late nineteenth-century theory of classical evolution, it was assumed that people outside of Western civilization were living representatives of a straight line evolutionary path, following pre-destined stages, that led from simple to complex and from savagery to civilization. Rather than recognizing that each culture has its own history, it incorporated other peoples into our history and, in advance, assigned them to a lowly place. The nonliterate peoples were called *primitive* because they were considered to be original, early, ancient, primeval. They were also considered to be childlike because childhood, too, was an early stage in human life. For the classical evolutionists, the mathematical ideas of traditional peoples were confined to number because, not only was it being taken for granted that traditional peoples were living ancestors, but it was also assumed that mathematics began with numeration. For example, according to one of the most highly regarded of the classical evolutionists, the beginning stages were assumed to be gestures using fingers and toes, then a few number words, then more number words, implicit bases of 5 preceded bases of 20 which preceded bases of 10, and so on. According to this story, the ordering of cultures on the intellectual scale began with native Australians, then native Americans, through Polynesians, Africans, and finally, of course, up to—and ending with—Western culture. Because they were in academic vogue, these assumptions and ideas soon became a part of the mathematical literature. Within this framework, numeral classifiers were mistaken to be evidence that traditional peoples lacked an abstract concept of number. And, carrying this even further, it was concluded that traditional peoples were only capable of concrete thought and not of abstraction or generalization. Then, in about 1910, another cultural theorist put forward the idea that there is a divided world of thought, that is, traditional peoples think differently—they are "prelogical" while we are logical. He, too, used numbers and systems of number words as an example to show that the minds of "primitives" do not function as ours do. His theories also strongly influenced historians of mathematics. Thus, there arose a confusion of early humans, traditional peoples, and developmental stages of Western children. This was combined with a characterization of traditional peoples as of lesser intelligence, incapable of analytic thought, without formal reasoning or logic, and capable only of concrete thought, not of abstraction or generalization. In the context of mathematics, this characterization was devastating.

During the past seventy years, much has been learned and these ideas about traditional peoples no longer reflect the thinking of cultural theorists. Because the ideas persisted in the mathematics literature, as well as in many other places, hosts of anthropologists have argued against them and worked to counter them. To do so, some have challenged the specific ethnographic examples of the early writers, and others have addressed the larger issue of logic and logical processes. Some anthropologists, for example, point to the linguistic evidence that none of the hundreds of languages studied so far lack the ability to handle the logical connectors *and*, *not*, *or*, *if ... then*, and *if and only if*. Others have demonstrated that the less formally structured logic of everyday discourse and everyday inference is, nonetheless, logical. Some philosophers, too, have contributed to the discussion by stating a belief in a universal "natural rationality." Among cognitive psychologists as well, the same issue has been addressed. Many have called into question earlier results that were based on tests used cross-culturally on adults and children alike, although developed in a Western context for developmental stages of Western children.

All of this is not to say that everyone thinks the same. The differences, however, are *not* in the ability to think abstractly or logically. They are in the subjects of thought, the cultural premises, and what situations call forth which thought processes.

We, in Western mathematics, for example, place a high value on decontextualization. Many others in our own culture, as well as in other cultures, do not. It is this difference, perhaps, more than anything else that has been confused with concrete thought and lack of abstraction. Let us take, for example, the biblical story in which King Solomon is faced with dividing one infant between two women who claim to be its mother. For this situation, we consider it wisdom that he chose to view the division *in context* and did not give one half to each. In fact, according to the story, he awarded the entire child to the woman who spoke out against the "objective" solution. In general, by decontextualization we mean the separation between form and content and context, and then dealing with the form as if it never had content or context. There is, in this approach, no concern for the contextual implications and ramifications of the formal manipulations. While often considered objective, the form that is imposed is neither necessary, universal, nor culture-free; the form itself is a selective creation whose very imposition has contextual ramifications. This approach is increasingly dominating our daily lives— it is, for example, the essence of computerized systems. But, concomitantly, there has been a marked increase in its questioning as people have become more aware of the social and environmental by-products. That is, what seemed successful when divorced from context or within a very narrow context, may not be that successful after all when viewed in its full context. Thus, the concern for context varies from situation to situation and from culture to culture, without having any bearing on the thought processes of which an individual or group is capable.

There are many distinctions between cultures: some produce food by hunting, others by farming, and others by fishing; some have many machines and some have few; some have systems of writing and some do not; some live in deserts, some in the Arctic tundra, and others live surrounded by water; and some are concerned with getting to Mars while others are concerned with getting to the Land of the Dead. All of these differences, and many others as well, affect the expression and content of mathematical ideas, but no single distinction will preclude those ideas or determine what they will be.

6 We distinguish our studies from traditional history of mathematics by using a different label, namely, *ethnomathematics*. The reasons for this are several. First of all, as we have said, we are concerned with the broader realm of mathematical ideas rather than the category called *mathematics*, which is limited in scope by its Western usage and connotations. Further, as we have also said, our interest extends to a wider spectrum of people within any culture. And, to us, *all* mathematical ideas, in traditional cultures or otherwise, need be viewed in cultural context. As such, we believe that traditional history has been limited to a singular expression within this worldwide panorama and, even when referring to other cultures, has done so using the viewpoint and bias of Western culture. All too often, even when including, for example, Chinese, Arabic, or Indian mathematical ideas, they are included only tangentially and with the Western expression as a measure or standard for comparison. Eventually, perhaps, there will be no need for the distinction between history of mathematics and ethnomathematics. For now, there is, because our studies require a different point of view that raises different questions and requires different understandings and different approaches.

To focus on traditional cultures, as we are doing here, mathematicians must draw upon, and interact with, disciplines they have rarely been involved with before. These disciplines are not the usual fields familiar to mathematicians. And, because until quite recently, the vast majority of traditional cultures had no systems of writing, they provided no written records that are the basic materials usual to history and historians. The written records that do exist are the observations and interpretations of others, not of or by the people whose ideas are being studied. Thus, who these others are, their knowledge of the language, the type and extent of their relationship to the culture being discussed, their understanding of cultural differences, and their theories of culture all become relevant to what they say. In short, to seek ideas embedded in cultures different from our own, it is necessary to become conversant with the work of ethnologists, linguists, and culture historians. My discussion of the kin relations of the Warlpiri, for example, drew on the findings of linguists, while the Malekula sand tracings and Maori game of mu torere were first recorded by ethnologists. Or, as in the case of the quipus and the Inca pottery designs, where there are material remains to be examined and analyzed, our study impinges on archaeology and archaeological interpretation. Thus, whether dealing with information gleaned from early travelers or from recent scholars or from artifacts, it is the anthropological fields that inform these studies.

Students of culture, however, have not been, and cannot be, equally familiar with or interested in all aspects of culture. Not being especially engaged with mathematics in their own culture, they rarely asked questions with mathematics in mind or were restricted in their view of what had mathematical import. Mathematicians, therefore, can bring to the ethnographic, linguistic, or archaeological record a perspective based on their broader and more particular understanding of mathematical ideas. Whether done as a collaborative effort or whether done by individuals grounded in one but conversant with the other, the study of the mathematical ideas of traditional peoples is at the interface of two fields of concern—mathematics and anthropology.

7   There are other somewhat similar sounding amalgams, such as ethnoscience, ethnobotany, or ethnoastronomy. Ethnoscience is decidedly different, as it is primarily a *method* for the study of traditional systems of knowledge and cognition. The method uses linguistic analysis to find the principles that underlie a culture's system of classification and categorization with regard to the natural world. Ethnobotany, on the other hand, is not that specifically or methodologically defined. In general, it focuses on the way a culture conceptualizes and interacts with plants in the environment. Similarly, ethnoastronomy deals with a culture's perceptions, beliefs, and activities related around celestial bodies. Ethnomathematics, in some sense, bears a resemblance to these. However, for our Western sciences called botany and astronomy, there are specific physical objects that give coherence to the disciplines. By extension, these same physical objects provide the domain of discourse for ethnobotany and ethnoastronomy. That is, the objects of study have physical reality, but how they are perceived or categorized and the beliefs that surround them can differ from culture to culture. For mathematics, however, there has been a long philosophical debate on the reality of the objects it studies. Is a square something that has external reality or it is it something only in our minds? Is the relationship between the length of the hypotenuse of a right triangle and the length of its sides a truth about reality that we have discovered, or is it something we have invented? Do prime numbers exist if we do not think about them? If we took the Platonic view that the objects of mathematics are there for the discovering, then the extension to ethnomathematics would revolve around these objects. According to the Platonists, mathematics is a science similar to, say, astronomy or botany. They believe that, while the objects are not physical and have no material substance, they *do* exist in some realm and are independent of our knowledge of them or interest in them. Since the late nineteenth century, there are many other philosophical stances. The majority of those who ponder about mathematics no longer believe in Platonism. Probably most practitioners of mathematics do not either, although they often speak and teach as if they do. To us, squares or right triangles or prime numbers are categories that we, in Western culture, have created (not discovered). These are our categories that others may or may not similarly create. The relationship between the length of the hypotenuse and lengths of the sides of a right triangle *is* an eternal truth, but that does not mean that any other culture need share the categories *triangle*, *right* triangle, *hypotenuse* of a right triangle, or *length* of a line segment. In fact, part of the reason we can be so certain about the relationships in the right triangle is that the objects are categorical ideals rather than real objects. So, as contrasted to astronomy and ethnoastronomy or botany and ethnobotany, Western mathematics and ethnomathematics are not linked to each other through objects that exist in the natural world. But, nonetheless, they are linked through phenomena of a different type, namely, mathematical ideas. The concept of number is a human universal. Also, all human groups live in space and, in one form or another, create a culturally shared ordering of that space as they communicate about it and function together within it. And, in addition to ordering space, humans impose order on much else in the natural and social world around them. The fact that cultures create categories, regardless of the categories they create, is evidence of that imposition of order. Categorization itself implies the seeking or establishing of relations. The creation and elaboration of orderly structures, interest in the relations within them and in their implications, pattern creation and pattern seeking, the creation and use of abstract models—these are the essence of mathematical ideas. Thus, while the unifying phenomena of ethnomathematics are different in kind from the objects of, say, ethnoastronomy or ethnobotany, it shares with them the goal of gaining insight into the intellectual realm of traditional cultures. And there will be some overlap between these studies when number, logic, or spatial forms are involved in explanatory or descriptive systems or, as in the case of Caroline Island navigation, when models are abstractions of the natural world.

8 There are today very few (if any) traditional cultures that have not been modified by the spread and overlay of a few dominant Western and Eastern cultures. Thus, while the traditional cultures persist into the present, by and large, for most of them, the ethnographic present, that time when they alone held full-sway, has become the past. There are, indeed, still strong and deep cultural differences but few, if any, people have not been affected in ways both large and small by the dominant cultures. If you look at Map 4 in Chapter 5, for example, you see some sites, such as the Solomon Islands, that became household words in the United States during World War II because they were the scenes of heavy fighting. Truk, in the Caroline Islands, was a Japanese naval base. Probably few of us were even aware that those islands had indigenous populations. And, in 1946, the word *Bikini* entered daily usage when the United States used this atoll in the Marshall Islands as a test site for an atomic bomb and as a name for a then radical style of women's bathing suits.

While I have focused here on ethnomathematics as part of a world-wide history, there are others who focus on its present and future ramifications. In particular, a concern for the interplay of culture and mathematical ideas has arisen among mathematics educators spread throughout the world. A critical issue is that, as it stands, much of mathematics education depends upon assumptions of Western culture and carries with it Western values. Those with other traditions are, as a result, often turned away by the subject or unsuccessful in learning it. And, for them, the process of learning mathematics, particularly when unsuccessful—but even when successful—can be personally debilitating as it detracts from and conflicts with their own cultural traditions, conceptual categories, and world view.

Some mathematics educators, such as those involved with government projects in administered territories such as Papua New Guinea, are interested in cultural differences that interfere with mathematical learning in its fully Western sense. Their goal is to be effective in overcoming these "deficits." However, the majority of other concerned educators are interested in modifying mathematics education so that it can effectively *build upon* and *reinforce* diverse cultural traditions. The issues are many and the issues are large. They involve distinguishing between mathematical ideas and their purely Western expression; distinguishing what is value-laden from what is not or, rather, selecting the values one wishes to transmit from those one does not; understanding cognitive processes and cognitive development across cultures; and developing curricula and creating teaching modes and materials appropriate to different cultural settings. In sum, the issues involve the goals of education in general, the place of mathematics education within these goals, and the means of carrying out these goals. These issues are not new. In fact, they are the same issues that educators have dealt with for hundreds of years. What is most significant, however, is that the impetus for addressing these issues anew has come from Latin American educators, African educators, and Native American educators. And, they bring to the discussions new and different viewpoints. While they have provided the impetus, interest in these issues is not limited to them but is shared by other individuals and groups in a wide variety of places including Australia, England, Guam, and Denmark. In some cases, such as the United States, the concern has been stimulated by the realization that our educational approaches have yet to come to grips with the fact that we ourselves are a multicultural society.

Ethnomathematics, in this broader perspective, is rooted in history but is much more than that. For it, in addition to mathematics and anthropology, social theory, educational theory, and cognitive studies all become involved.

9 One of the major differences between the twenty-first century and the twentieth century will be the degree of intermingling of people of different cultures. Every day, more aspects of the lives of all of us are being affected by this. Just as in any other realm, participants in the world of mathematics cannot be adequately prepared for the future without an understanding of the past and of the present.

Western mathematical ideas have changed through time; so, too, have our philosophies and histories of mathematics. Not the least of these changes has been the definition and redefinition of the boundaries of mathematics. It is important to once again revise our philosophy and history in recognition of the fact that mathematical ideas are cultural expressions and that our Western ideas are intimately linked with Western culture. And, moving beyond our own Western mathematics, a broader global ongoing history must acknowledge and include the ideas of other cultures.

Notes

2. Although their concerns do not extend beyond Western culture, there are a number of historians of mathematics who see Western mathematics as an intimate part of its cultural and social milieu. "The centrality of mathematics in the history of Western thought," J. V. Grabiner, *Mathematics Magazine*, 61 (1988) 220–230, is an excellent presentation of the interrelationship of Western mathematics and Western thought. R. L. Wilder (*Mathematics as a Cultural System*, Pergamon Press, New York, 1981) used his understanding of culture to describe the processes of mathematical development in the West. In essence, he views professional mathematicians as a subculture of Western culture. In "The interactions of mathematics and society in history: some exploratory remarks," *Historia Mathematica*, 4 (1977) 7–30, H. J. M. Bos and H. Mehrtens argue for doing the social history of mathematics. D. J. Struik in "The sociology of mathematics revisited: A personal note," *Science & Society*, L (1986) 280–299 presents an overview of writings that link Western mathematics and society. The work of S. Restivo, a sociologist of mathematics, is also recommended. See, for example, his "Representations and the sociology of mathematical knowledge," *Les Savoirs Dans Les Pratiques Quotidiennes*, C. Belisle and B. Schiele, eds., Editions du Centre National de la Recherche Scientifique, Paris, 1984, pp. 66–93.

5. For a succinct discussion of the theory of classical evolution see "Classical evolution" by R. L. Carneiro, pp. 57–121 in *Main Currents in Cultural Anthropology*, edited by R. Naroll and F. Naroll, Prentice-Hall, Englewood Cliffs, New Jersey, 1973 or M. J. Herskovits' "A genealogy of ethnological theory," pp. 403–415 in *Context and Meaning in Cultural Anthropology*, M. E. Spiro, ed., The Free Press, New York, 1965. It was the prominent classical evolutionist E. B. Tylor who devoted an entire chapter of his *Primitive Culture*, volume 1, Estes and Lauriat, Boston, 1874, to "The Art of Counting" (pp. 240–272). His ideas pervade the book *The Number Concept: Its Origin and Development*, written in 1896 by the mathematician L. L. Conant (MacMillan and Co., New York). This book has been quite influential and continues to be read and cited. For example, its first chapter is reprinted as "Counting" in the widely read *The World of Mathematics*, volume 1, J. R. Newman, ed., Simon & Schuster, New York, 1956, pp. 432–441, and, among other places, it is cited by C. B. Boyer in *A History of Mathematics*, John Wiley and Sons, New York, 1968; H. W. Eves, *An Introduction to the History of Mathematics*, 5th ed., Saunders College Publishing, New York, 1982 (first edition 1953); and G. Ifrah, *Histoire Universelle des Chiffres*, Seghers, Paris, 1981 (English translation *From One to Zero: A Universal History of Numbers*, Viking, New York, 1985). The notion that nonliterate people are "prelogical" was developed by L. Lévy-Bruhl, in *How Natives Think*, George Allen and Unwin Ltd., London, 1926 (original French edition 1910). His ideas entered the mathematics literature with the original 1912 edition of L. Brunschvicg's *Les Étapes de le Philosophie Mathématique*, A. Blanchard, Paris, reprinted 1972. They are found with force in K. Menninger's *Zahlwort und Ziffer*, Vandenhoeck and Ruprecht, Göttingen, 1957 (*Number Words and Number Symbols*, MIT Press, Cambridge, Mass., 1969 is an English translation without bibliography of a 1958 edition), and still a generation later in, for example, the book by Ifrah cited above.

A more detailed discussion of the pervasive characterization of traditional peoples in the mathematics literature is in "Ethnomathematics," M. Ascher and R. Ascher, *History of Science*, 24 (1986) 125–144. That article also includes more about writings in anthropology, philosophy, and cognitive psychology that counter this view. Two additional articles that specifically discuss cultural differences in contextualization and abstract formalism are "Contextualization and differentiation in cross-cultural cognition," J. P. Denny, *Cognitive Science Memorandum No. 28*, Centre for Cognitive Science, University of Western Ontario, London, Ontario, 1986, and S. Buck-Morss, "Socio-economic bias in Piaget's theory and its implications for cross-cultural studies," *Human Development*, 18 (1975) 35–49. Particularly recommended is *Cultural Context of Learning and Thinking*, M. Cole, J. Gay, J. A. Glick, and D. W. Sharp, Basic Books, New York, 1971.

The biblical story is in Kings I, Chapter 3, verses 16–28. "Applied mathematics as a social contract," P. J. Davis, *Mathematics Magazine*, 61 (1988) 139–147 is an excellent statement of the need to question the growing effects of (what he calls) "mathematization."

7. For an overview of ethnoscience and various ethno studies, see "Native knowledge in the Americas," C. S. Kidwell, *Osiris*, 1 (1985) 209–228. The rise of ethnoastronomy, in particular, is documented in "Archaeoastronomy and ethnoastronomy so far," E. C. Baity, *Current Anthropology*, 14 (1973) 389–431. The article is followed by about twenty pages of comments written by nineteen other scholars.

An imaginative and interesting approach to the relationship of mathematics to reality is A. Rényi's *Dialogues on Mathematics*, Holden-Day, San Francisco, 1967. Chapter 7 (pp. 318–411) of *The Mathematical Experience*, P. J. Davis and R. Hersh, Houghton Mifflin Company, Boston, 1981, is a highly readable discussion of some of the same philosophical issues raised here. Also, for an introduction to the philosophy of modern mathematics, see "The three crises in mathematics: logicism, intuitionism, and formalism," E. Snapper, *Mathematics Magazine*, 52 (1979) 207–216 and reprinted in *Mathematics: People, Problems, Results*, volume 2, D. M. Campbell and J. C. Higgins, eds., Wadsworth International, Belmont, California, 1984, pp. 183–193. As noted in the editors' preface to the article, it relates the philosophies to their views about the nature of the "abstract entities" studied by mathematicians.

8. The Indigenous Mathematics Project in Papua New Guinea is discussed by D. F. Lancy in *Cross-Cultural Studies in Cognition and Mathematics*, Academic Press, New York, 1983.

Claudia Zaslavsky's *Africa Counts*, Prindle, Weber & Schmidt, Boston, 1973, served as an inspiration for many educators. U. D'Ambrosio (Brazil) is prominent in the multicultural educational movement. Among his numerous writings are "Ethnomathematics and its place in the history and pedagogy of mathematics," *For the Learning of Mathematics*, 5 (1985) 44–48; *Socio-Cultural Bases for Mathematics Education*, UNICAMP, Campinas, Brazil, 1985; and "Ethnomathematics: A research program in the history of ideas and of cognition," *ISGEm Newsletter*, 4 (1988) 5–8. The latter two also provide references to D'Ambrosio's other writings. Involved African scholars include A. Djebbar (Algeria), P. Gerdes (Mozambique), and L. Shirley (Nigeria). As part of the African Mathematical Union, in 1986 the Commission on the History of Mathematics in Africa (AMUCHMA) was created. Its first newsletter appeared in 1987. The *AMUCHMA Newsletter* is an important resource; it carries names and addresses of interested scholars and references to their work, as well as notices of meetings and associated activities. The Native American Science Education Association (NASEA) was created in 1982. Their projects include the development and use of elementary and secondary school mathematics curricula in which the lessons and materials are related to the experience and culture of Native Americans. Extensive mathematics teaching materials have been created by C. G. Moore (Northern Arizona University). The NASEA also publishes an informative newsletter, *Kui Tatk*.

In 1985, at the annual meeting of the National Council of Teachers of Mathematics, a group of educators formed the International Study Group on Ethnomathematics (ISGEm). The ISGEm serves as a focal point by publicizing, organizing, or participating in sessions at national and international meetings (e.g., NCTM, IACME, ICME) and by publishing a newsletter. Their newsletter is probably the best source of up-to-date information about this increasingly active educational movement. Another group whose interest in ethnomathematics is part of a broader interest is the International Study Group on the Relations Between History and Pedagogy of Mathematics. Their activities and newsletter are also important resources.

i n d e x